3X Your Law Firm Cases

Kenneth Sullender

3X Your Law Firm Cases

by Getting Your Internet Marketing Right

Paperback ISBN: 978-1-952098-48-2
eBook ISBN: 978-1-952098-49-9

Printed in the United States of America.

To contact the author, please send email to:

internetmpros@gmail.com

CONTENTS

ONE

Your Online Marketing Plan

(Website, SEO, PPC, Pay-per-lead services, etc.)

Congratulations on your purchase of "How to Triple your law firm cases by Getting your Internet Marketing Right," a complete overview of what it takes to maximize your opportunities online in terms of Leads, Calls and Revenue.

There are a number of channels/mediums to consider for your law firm when you look at the Online Marketing space. Whatever specialty, this book has the plan for you.

At first glance, considering all the marketing options available in your online marketing playbook might be overwhelming. These include Search Engines (Organic, Maps, Pay-Per-Click), Social Media (Facebook, Twitter, LinkedIn), Paid Online Directory Listings (BOTW, FourSquare, Yelp.com, etc.) and Paid Online Lead Services.

To maximize your lead flow from the Internet, you need to develop a PLAN, which covers each of these online marketing opportunities. The purpose of this book is to outline a plan that will transform you from an online marketing novice to the dominant player in your area.

Throughout this book, we lay the foundation to:

1. Map out your online marketing plan (Website, SEO, PPC, Pay-Per-Lead services, etc.)
2. Start with the fundamentals (Market, Message, Media) before jumping headfirst into your Internet Marketing Strategy.
3. Setup your website.
4. Understand how search engines work, and learn the differences between the paid, organic and map listings.
5. Optimize with Search Engine Optimization - How to optimize your website with keywords that are most important for your particular specialty.
    - How to conduct keyword research.
    - Our list of the most commonly searched keywords broken down for law firms.
    - How to achieve the maximum result by mapping out the pages which should be included on your law firm website.
    - How to optimize your website for ranking in the organic listings on major search engines.

- How to improve your website's visibility so you can rank on page one for your most important keywords.
- List of link building techniques and strategies that are proven to enhance rankings even in the post Penguin and Panda Era.
- Content marketing strategies for maintaining relevance in your market.
- Optimize Google Maps - How to get ranked on the Google Map in your geographic area for 'near me' searches.
- The fundamentals of Google Maps ranking (NAP, Citations, Consistency and Reviews).
- How to establish a strong name, address, and phone number profile.
- How to properly claim and optimize your Google My Business Local Listing.
- How to develop authority for your map listing via citation development.
- List of the top citation sources for law firms, according to the industry standards.
- How to get real reviews from your customers in your true service area.
    - Sample Review Card
    - Sample Review Request Email.
    - Sample Review Us landing page for your website.

6. Understand Website Conversion Fundamentals - How to ensure that your website converts visitors into leads in the form of calls and web submissions.
7. Understand Mobile Optimization - How to optimize your website for mobile visitors.
8. Utilize Social Media Marketing - How to utilize Social Media (Facebook, Twitter, LinkedIn and other social platforms for maximum effect for your law firm.
9. Use Video Marketing - How to tap into the power of YouTube and other video sharing websites to enhance your visibility and drive better conversion.
10. Leverage email marketing tools (Constant Contact, Mail Chimp, etc.) to connect with your customers on a deeper level, receive more reviews, get more social media connections and ultimately get repeat and referral business.
11. Understand and capitalize on Paid Online Advertising opportunities.
12. Use Pay-Per-Click Marketing (Google AdWords and Bing Search) - How to maximize the profitability of your Pay-Per-Click Marketing efforts.
    - Why PPC should be part of your overall online marketing strategy.
    - Why most PPC campaigns fail.
    - Understanding the Google AdWords Auction process.

- How to configure and manage your Pay-Per-Click campaign for maximum ROI.

13. Use Paid Online Directories - What paid online directories should you consider advertising in (FindLaw, Avvo, BOTW, Spoke.com, Yahoo, Yelp, Foursquare, Yellowpages etc.).

14. Manage Pay-Per-Lead and Lead Services - How to properly manage Pay-Per-lead services for maximum return and long-term gains.

15. Sample lead follow up email sequence.

16. Track, Measure and Quantify - How to track your online marketing plan to ensure your investment is generating a strong return.

When it comes to Internet marketing for your law firm, there are several avenues to explore. In this chapter, we will briefly touch on the various Internet marketing channels that are available, and then go into more detail throughout the book.

This chapter serves as your "Marketing Plan" and roadmap going forward.

Online Marketing Channels

1. Search Engine Optimization (Organic Listings and Map Listings).
2. Search Engine Marketing/PPC on Google AdWords and Bing Search Network.
3. Social Media Marketing (Facebook, Twitter, LinkedIn).
4. Video Marketing.
5. Email Marketing.
6. Paid Directory Marketing (BOTW, Spoke.com, Yahoo, Yelp, Foursquare, Yellow Pages etc.)
7. Paid Lead Services (Emfluence.com, Fuellead.com, Intellibright.com, etc.)

Search Engine Optimization

Search Engine Optimization (SEO) is the process of increasing your Law Fim's visibility on major search engines (Google, Yahoo, Bing, etc.) in the organic, non-paid listings as consumers are searching for your specialty or services.

There are three very critical components of Search Engine Marketing. The three components are:

- Paid Listings – The area along the top and side that advertisers can bid on and pay for in order to obtain decent placement in the search engines.
- Organic Listings – The area in the body of the Search Engine Results page.
- Map Listings – These are the listings which come up beneath the paid listings and above the organic listings in a number of searches.

Search Engine Optimization involves getting your website to show up in the Organic and Map Listings. These listings account for a majority of the search volume. As depicted in the illustration below, more than 78% of searchers click on the Organic (non-paid listings) rather than the paid listings.

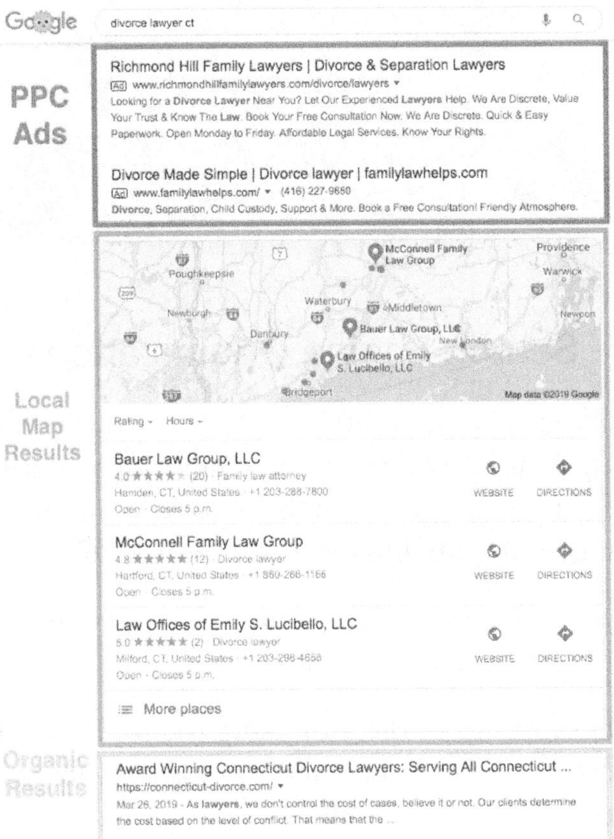

When most people think "Internet Marketing," they think Search Engine Optimization. However, you will begin to see that SEO is only a small piece of the MUCH BIGGER "Internet Marketing" puzzle for law firm owners.

Search Engine Marketing / Pay-Per-Click

Now that we have discussed SEO, let's talk about Search Engine Marketing or PPC (Pay-Per-Click). Google, Yahoo and Bing all have paid programs that allow you to BUY listings associated with your keywords to be placed in designated areas of their sites.

There are three really important benefits of PPC:

- Your keyword listings will appear on search engines almost immediately.
- You only have to pay when someone actually clicks on your listing – hence the term Pay-Per-Click Marketing.
- You can get your ad to show up on national terms in the areas/cities in which you operate.

PPC Marketing works on an Auction system similar to that of eBay. You simply choose your keywords and propose a bid of what you would be willing to pay for each click.

There are several factors that determine placement, which will be discussed in detail in the PPC for law firms chapter. But, in the broadest sense, the one who is willing to pay the most per click will be rewarded the top position in the search engines, while the second-most will be in the second position, etc.

PPC Marketing is a great way to get your law firm's website to appear at the top of the search engines right away, driving qualified traffic to your website.

Social Media Marketing

There is a lot of buzz around Social Media (Facebook, Twitter, Instagram, LinkedIn, YouTube), but how can it be utilized by a small law firm which is just starting out? How can you use social media to grow your law firm? Just look at the staggering statistics for Facebook:

- More than 500 million active users.
- 50% of active users log-on to Facebook on any given day.
- Average user has 130 connections.
- People spend over 700 billion minutes per month on Facebook.

So, how can you employ this amazing tool to grow your law firm? Use it to connect with your personal sphere of influence, past and new customers. By doing so you can solidify and maintain existing relationships, remain top-of-mind, and ultimately increase repeat and referral business.

Video Marketing

Did you know YouTube is the second-most used search engine on the market? Would you guess it is even ahead of Bing and Yahoo? It's true!

Millions of people conduct YouTube searches on a daily basis. Most business owners are so focused on SEO they completely neglect the opportunities that video and YouTube provide.

By implementing a Video Marketing Strategy for your law firm, you can get additional placement in search results for your keywords, enhance the effectiveness of your SEO efforts and improve visitor conversion.

Email Marketing

Similar to Social Media Marketing, email marketing is a great way to remain top-of-mind with your clients and increase repeat business and referrals. Compared to direct mail and newsletters, email marketing is by far the most cost-effective means to communicate with your clients.

As we will discuss in the Email Marketing for law firms chapter, we feel email marketing can be used to effectively draw your clients into your social media world.

Paid Directory Marketing

There are a number of Online Directories that are important for law firms:

- BOTW
- Spoke.com
- Yahoo
- Yelp
- Foursquare
- YellowPages
- FindLaw
- Avvo
- Justia

Paid Lead Service Sites

There is an array of services that will sell you leads on a "pay-per-lead" basis or a flat monthly fee.

- Emfluence.com
- Fuellead.com
- Intellibright.com

While these leads tend to go to a number of different providers and will be less qualified than other sources, these Pay-Per-Lead services can be a profitable online marketing channel if executed correctly.

Now that you have an understanding of each of the Internet marketing channels available, in the following chapters we will discuss how you can leverage them to connect with new clients and grow your law firm.

Where to Start?

With such a large amount of Internet marketing channels, where should you start? I firmly believe that over time, you should be appropriating each of these online marketing opportunities.

However, you must first begin with the foundation - your website, organic rankings and social media/email. You should start looking at the various paid marketing opportunities when your website is set up correctly, ranking on search engines for your most important keywords in the organic, non-paid listings and you are actively engaging in social media activity.

We have found that the biggest and most impactful opportunity is getting ranked organically (in the non-paid listings). You may then leverage the additional profits in paid marketing to further augment your growth.

Once you are ranking well organically and things are firing on all cylinders, then you can start to run a well-managed Pay-Per-Click Campaign and explore paid online directory listings on Yelp, Foursquare, etc.

Next, let's look at the fundamentals of your overall marketing strategy before pressing forward into full implementation.

TWO

Start with the fundamentals

(Market, Message, Media) before jumping headfirst into your Internet marketing strategy

Before we delve into Internet Marketing, SEO, and Social Media Marketing, I want to be sure we have built a strong marketing foundation.

As I talk with various law firms across the United States, I have come to the realization the vast majority tend to skip straight past the basic fundamentals of your marketing strategy and dive headfirst into tactics (Pay-Per-Click adverting, SEO, Social Media, etc.).

So, what do I mean when I say "Fundamentals"? All marketing has 3 core components:

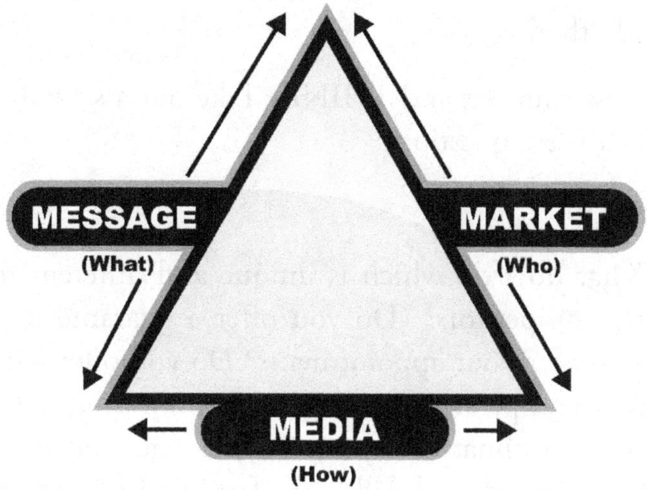

- Message (what)
- Market (who)
- Media (how)

You have to have a unique "Message" (who you are, what you do, what makes you unique, and why someone should hire you rather than another law firm offering the same service), a specifically defined "Market" (who you sell to and who your best clients are), and then look at "Media" (where you can reach those best clients). The tactics (Pay-Per-Click, SEO, Social Media, Direct Mail, etc.) fall into the "Media" category.

If you focus solely on the Media or Tactics, you will likely fail regardless of how well-selected that Media is. With that being said, you need to scale back to the fundamentals. Invest the time and energy in fleshing out your "Message" and figuring out who your "Market" is. By doing so, ALL of your Media choices will be vastly more effective. How can you do that?

Spend a few minutes and THINK. Take out a scratch pad and answer these questions:

Message:

- What do I do which is unique and different from my competitors? (Do you offer a guaranteed time frame for your appointments? Do you offer written estimates prior to starting work, promising to stand by that estimate? Maybe you offer a guarantee for all of your work and will look after any issues within a one-year period of time after the project is complete. Perhaps you offer a free consultation to showcase the services you have to offer to potential customers.)
- If you think about the psychology of a client, what concerns or apprehensions do you think they have about hiring the services of a law firm? "Do they have the experience for me to win", "How much will all this cost me", "How do I know this law firm is the best"?
- How can you address your clients common concerns in a unique way?

Market:

- Who is my ideal client? (Please realize not everyone resides in your city nor within a 25-mile radius of your office). You need to be clear about the audience that you are looking to attract.

- Look at your last 25 clients and evaluate who spent the most money, who had the highest profit margins, and who was genuinely pleased with your service. What are the unique characteristics of those good clients? Do they live in a particular area of town? Do they have a higher income level? How did they hear about your service offerings?
- Start to define who your ideal client is so you can put a marketing plan in place to attract similar clients.
- Once you have fleshed out your Message and your Market, then you can start to think about Media. In order to determine what media will be most effective for you, you need to think about where you can reach your IDEAL client.
- Clearly, the Internet is a great "media" for connecting with your ideal client who is proactively in the market for your services. Throughout the remainder of this book, we will be explaining the various Internet marketing channels and how you can use them to connect with your ideal client.

Remember, you need to start with the FUNDAMENTALS (Message, Market and Media) before running headstrong into any marketing.

THREE

How To Setup Your Website

This chapter is all about how to setup your law firm website. We are going to cover a lot of details as they relate to SEO, Google Maps Optimization, Pay-Per-Click Marketing, etc.

However, without a properly designed and functioning website, those efforts will be put to waste. Before you can or even should begin exploring those options, you must have your website up and running.

Formats

Let's talk about website formats and the different options available to you when you are ready to start.

1. HTML Site – There are basic HTML pages and/or individual pages that can be incorporated into a website.

This is how almost all websites were built several years ago. They had multiple pages hyper-linked together.

2. Template Based Site Builders - Site builders, which you can obtain through providers such as Go Daddy, Website Tonight, and 1&1 are turn-key. You buy your domain and set up your website. I have found this type to be quite a bit less than ideal because you don't have a lot of controller flexibility. But, there are still a lot of sites in this format.

3. CMS Systems - Content Management Systems, like WordPress, Joomla, Drupal. I'm sure there are many others but these are the big ones.

Given my experience with websites, a content management system (CMS) is ideal for a law firm. I say that because you have scalability. In any of these platforms, you have the ability to change your navigation on the fly, add as many pages as you need and easily scale out your site.

If you have your website built in WebsiteByTonight or in HTML format with graphics behind the website, and you wanted to add a new section, you would have to start from scratch.

You would have to go back to the graphics and modify all of the pages in order to add the new section to your navigational structure. With a CMS, everything is built behind code allowing the ability to apply easy edits and to add multiple pages.

As you will see in the search engine optimization section of the book, you will have the ability to have a page for each one of your services and each city in which you operate.

A CMS allows you to create your pages in a scalable format without having to mess around with the graphics or do anything that is difficult to control. Also, it is easy to access, modify, and update.

Using formats like WordPress and Joomla, you may access the back-end administrative area at yourcompany.com/ login.

After entering your username and password, you will find there is a very easy to edit system with pages and posts which function similarly to Microsoft Word.

You can input text, import images and press "save", forcing all new edits to be updated on your live website. It is easier than it looks and is very search engine friendly.

Content Management Systems have intelligently structured linking between pages and content, making it extremely search engine friendly. We have found this method tends to be better than regular HTML or Website Tonight options.

In a lot of cases, a blog is going to be automatically bolted onto a CMS based website providing you with a section where you may feed updates. In the SEO chapter, we cover the importance of creating consistent updates and blogging regularly.

Another benefit of content management systems is being provided with a variety of plug-ins you can choose to incorporate on your website. You can easily pull in your social media feeds, YouTube Videos and check-ins.

You may also syndicate your website to automatically post any new updates to your social media profiles. You can add map integration where people can click to either get instructions or view a map to find out the areas, which are served by your law firm. There is a surplus of features available within a CMS that you can't necessarily do with a non-CMS type option.

Whether you are looking to build a website from the ground up, you are just getting started, or you feel like you simply need a redesign, I highly suggest that you do so in CMS, ideally in WordPress.

WordPress is a fantastic platform and very easy to use. It's the most adopted website platform available with many developers using it. It's constantly being updated and improved and I have found it to work very well for different law firm practices.

You have my stamp of approval to go out and build your website on a WordPress platform.

What Should Your Law Firm Website Include?

So, what pages should your website have? What navigation structure should you create? Depending on your law firm, you will need to showcase different things. For most Firms, though, the basics should be:

1. Home
2. About Us
3. Our Services
4. Our Service Area (You will understand what I mean once you read the SEO Chapter)
5. Online Specials or Coupons
6. Reviews and Testimonials
7. Testimonials
8. Client Q&A
9. Blog
10. Contact Us

These are the core pages. Within "About Us," you might incorporate a drop- down menu for subcategories including

"Meet the Team," "Why Choose Our law firm," etc. I think that's very powerful.

You want to be able to drive people back to a "Why Choose Us" section, and, in some cases, if you are having issues recruiting and retaining good quality talent, you might want to have a "Careers" page under the "About Us" navigation, where a visitor can go and fill out an application and learn more about your law firm.

Within "Our Services," you want to have the ability to list a drop-down listing the types of services or areas of practice that you offer. We discuss this to a great extent in the SEO chapter.

You want to have landing pages for each one of your services because they are going to be optimized with different keyword combinations.

If you have a law firm where you make consults to homes or businesses, a "Service Area" section will give you the ability to show a heat map of all of the locations that your team goes to, as well as a drop down menu that lists the sub-cities in which you operate within your market.

A "Reviews and Testimonials" page will provide a section to showcase what your clients are saying about you in text or video form.

You can also pull in reviews from sites such as Google Maps, Angie's List, and Yelp. Finally, of course, you will need a "Contact Us" page where web visitors have your general contact information.

These are the core things you should have on your website.

A Clear Description of Who You Are

A visitor who stumbles upon your website shouldn't have to do a thorough investigation to figure out who you are and what exactly you do.

This means it's important to clearly mention your law firm name and sum up your products or services 'above the fold' section of your website. A clear and specific description will attract the visitor's attention immediately - within two to three seconds - and encourage them to spend time on your website.

Your Primary Contact Details

Outside of your navigational structure, what else should your website have? What other elements are going to help with conversion?

Well, you should always provide a primary phone number on every page of your website, in the upper right-hand corner. When somebody visits a page, their eyes are naturally drawn to the top section of the website where they can see the logo and the phone number.

People tend to expect that phone number will be somewhere in this location. It is ideal to have a prominent phone number, telling them to "call you now" for service in that section.

An Obvious Call to Action

I believe law firm websites should always make a web form available from which a client can easily request a quote.

Bear in mind that every visitor to your website is in a different situation and frame of mind. You may have someone that's on his or her phone or just leisurely looking to contact you for your Firm's services and is able to simply pick up the phone and call you.

On the other hand, somebody who's in a work environment may not have the ability to stop what he or she are doing and make a phone call without drawing attention from his or her coworkers. However, they may be able to browse around online to find out what options are available.

Your potential clients may reach your website and be torn between making a call right at that moment, just scheduling the appointment, or wanting to have someone from your team contact them.

Make it easy for them to enter their information into a web form where they can provide their name, phone number, email address, and a note detailing their requests that they can send through online. It makes it easier and doesn't create any pressure. You should also incorporate web chat. This allows for a more discreet opportunity for a potential client to contact your firm. There are plans that allow your team to directly manage the chat request or you can outsource that to experienced service providers.

Social Media Links

You also want to provide links to your social media profiles. Link to Facebook, Twitter and LinkedIn so clients can easily jump off, engage with you on social media, see what you're doing and be able to press that important "like," "follow" or "subscribe" button. It helps create a sense of authenticity when your clients get to see your social media content.

Client Testimonials

Have a direct link, which drives visitors to your online reviews and testimonials that we discussed previously.

You should also post your credentials either in the sidebar or in the header graphic, proving, for example, that you're BBB-accredited or a member of the local chamber of commerce or Legal association.

This allows potential clients to rest assured that you are a credible law firm, you're involved in the community and that you're less apt to provide them with poor service. They'll feel more comfortable doing business with you.

You definitely need to have your law firm name, address and phone number on every page of your website.

It is not critical that you list your address on each page because it will not be a determining factor in whether or not clients call you, but as I will explain in the Google Maps optimization chapter, having name, address and phone number consistency is critical for ranking on Google Maps.

It is a great strategy to have your name, address and phone number referenced on your website, ideally in the footer section. You need to have that contact information on all of your pages including the Contact Us page, of course.

Authentic Images

It's extremely important that you infuse personality into your website. By personality, I'm referring to authentic photos and videos.

Showcase your law firm, feature yourself, the business owner, and the people who work in the firm: the office team, the paralegals, etc.

Showcase the office itself, the conference room(s), and facility if you have them. Don't use stock photography, but authentic imagery. This gives the client the chance to get to know, like and trust you, before they even pick up the phone. I've seen this tactic prove itself time and time again.

Say a potential client visited two different sites for a similar law firm. One of them is generic; here's the same image he or she has seen before of the same law firm with the same team on multiple other websites.

Now the other website highlights a genuine picture of the actual firm, the team and facilities. This authentic page converts 10 to 1. You must let your real personality reflect on the website.

You must also craft messaging that explains why they should choose your aw firm. Why should someone choose you over the competition?

Pull them down a path where they can start to learn more about why you are their best option. Where they can see your online reviews, and if they're kind of on the fence, where they can quickly locate some special offers and incentives that will drive action.

This will get them to contact you right away, as opposed to continuing to browse the web for someone else.

Mobile Website

The other major thing you want to think about, from the conversion perspective, is having a mobile-ready version of your website.

More and more people are accessing the Internet via smart phones such as iPhones, Androids, and ipads. Make sure the mobile version of your site isn't the same as your regular site. It must be mobile friendly and pass the Google Core Vital requirements. You can perform a fee check of your law firm's website by visiting: https://developers.google.com/speed/pagespeed/insights/

It should be condensed, fitting their screen and giving them just the information that they need. It should integrate with their phone so all they have to do is press a button to call you.

People who are searching or accessing your website from a mobile device are in a different state of mind than the people that are browsing and finding you on a computer. Make it easy for them to get the information they need and to get in touch with you.

FOUR

Understanding HOW Search Engines Work

And the Differences Between the Paid, Organic and Map Listings

In this section, we want to take a few minutes to demystify the search engines and break down the anatomy of the Search Engine Results Page. By understanding how each component works, you can formulate a strategy to maximize your results.

There are three core components of the Search Engines Results page:

1. Paid/PPC Listings
2. Map Listings
3. Organic Listings

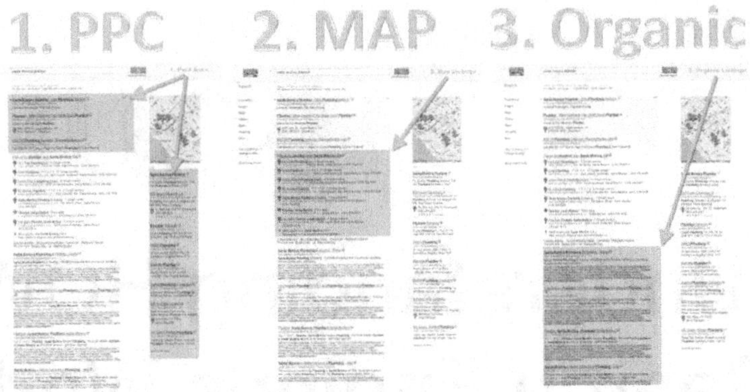

Paid/PPC Listings – In the paid section of the search engines you can select keywords which are relevant to your law firm and specialty, and then pay to be listed amongst the search results. The reason it is referred to as PPC or Pay-Per-Click[1] is because rather than paying a flat monthly or daily fee for placement, you simply pay each time someone clicks on the link.

Map Listings – The map listings have become very important because they are the first thing that comes up in search results for most locally based searches. If someone searches for some particular specialty or service in your area, chances are the map listings will be the first thing they look at. Unlike the paid section of the search engine, you can't buy your way into the Map Listings. You must earn it. Once you do, there is no per-click cost associated with being in this section of the search engine.

Organic Listings – The organic/natural section of the Search Engine Results page appears directly beneath the Map Listings in many local searches but appears directly beneath the Paid Listings in the absence of the Map Listings (the Map Section only shows up in specific local searches). Similar to the Map Listings, you can't pay your way into this section of the search engines and there is no per-click cost associated with it.

Now that you understand the three major components of the Search Engine Results and the differences between Paid Listings, Map Listings and Organic Listings you might wonder… "What section is the most important?"

This is a question that we receive from a large number of law firms every day.

The fact is all three components are important, and each should have a place in your online marketing program because you want to show up as often as possible when someone is searching for your service offerings in your area.

Return On Investment

With that said, assuming you are operating on a limited budget and need to make each marketing dollar count, you need to focus your investment on the sections that are going to drive the strongest Return On Investment(ROI).

Research indicates that the vast majority of the population looks directly at the Organic and Map Listings when conducting a search, and their eyes simply glance over the Paid Listings (as illustrated by the images below).

Google search: atlanta personal injury lawyer

Desktop search results | Mobile search results

So, if you are operating on a limited budget and need to get the best bang for your buck, start by focusing your efforts on the area which gets the most clicks at the lowest cost. We have found placement in the Organic and Map section on the Search Engines drive a SIGNIFIGANTLY higher Return On Investment than Pay-Per-Click Marketing.

Begin with the Organic Listings and then, as you increase your profits, you can start to shift those dollars into a proactive Pay-Per-Click Marketing effort.

In the next chapter, we will start to look at Search Engine Optimization and how to optimize your website to rank in the organic listings (non-paid) for the most important keywords in your field.

How Do Search Engines Work?

It is also important to understand how search engines work. This includes the process of crawling and indexing, plus the concept of page rank as well.

Search engines work by crawling billions of web pages using their own web crawlers or web spiders. These web crawlers are also known as **Search engine bots**[2].

Understanding Search Engine Index

Once a webpage is discovered by search engine, they are added into a search engine data structure which is called 'index' or 'indexing'. Search engine index includes all the crawled web URLs along with several important key elements about the content of each web URL such as:

- The keywords
- Type of content
- Uniqueness of the page
- User engagement with the page

Understanding Search Engine Algorithm

Search engine algorithm aims to display a relevant set of high-quality search results that will fulfill the user's search query as quickly as possible.

What Happens When a Search Query is Entered?

When a search query is entered into the search engine by a potential user, the search engine tries to identify all the pages which are deemed relevant.

During this process, the search engine uses a special algorithm to hierarchically rank the most relevant pages into a set of results. The algorithm which is used to rank the most relevant web pages differs for each search engine.

For example, a web page that ranks on the top for a search query in Google may not rank highly for the same query in Bing.

Mentioned below are a few elements search engines use to return the results.

- Search query
- Location
- Language detected
- Previous search history
- Device from which the search query was entered

Sources and References:

1. https://en.wikipedia.org/wiki/Pay-per-click
2. https://en.wikipedia.org/wiki/Web_crawler

FIVE

Search Engine Optimization –

How to optimize your website for the keywords that are most important for your particular business

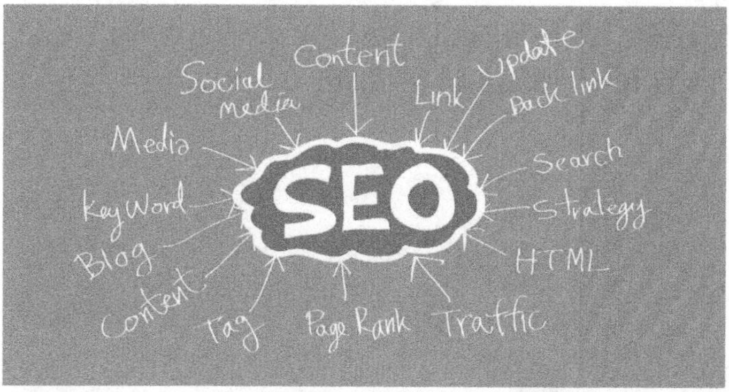

Getting your law firm listed in the organic section (non-paid listings) of the search engines comes down to two core factors:

- Having the proper on-page optimization so Google knows what you do and the general area you serve. This allows it to put in the index for the right keywords. You do this by having pages for each of your services and then optimizing them for specific legal keyword combinations (Ex. Your City + main service, Your City + service 2, Your City + service 3, etc.).
- Creating enough authority and transparency so Google ranks you on page one (rather than page ten) for those specific keywords. Ultimately, it comes down to having credible inbound links and citations from other websites to your website and its sub-pages. He who has the most credible inbound links, citations and reviews will be the most successful.

Throughout the course of this chapter I provide specific how-to information on exactly what pages to add to your business website - and why. I also discuss what you can do to improve your authority/transparency in Google's eyes so your website ranks on page one for the keywords which are most important to your law firm.

Before you start creating pages and trying to do the "on-page optimization" work, you need to be clear on the most commonly searched keywords relative to the services you offer.

By understanding the keywords, you can be sure to optimize your website for the words that will actually drive qualified traffic to your site. One needs to conduct a detailed research of the market and the requirements that potential clients have in order to find the optimal keywords which will help you bring in more clients.

Given that different organizations might be working in different industries, it is imperative businesses learn the methodology behind selecting the most relevant keywords for their services. We have provided an overview of how to conduct keyword research.

How to Conduct Keyword Research

To determine what your clients are searching for when they need your services, here are a number of tools that can be used to conduct keyword research. Some are free of charge while others have a monthly cost associated with them. Some of the better keyword research tools include Wordstream, Google AdWords Keyword Tool and SEM Rush.

For the purposes of this book, we have developed instructions based on the free Google AdWords Keyword tool. To use Google AdWords Keyword tool, you'll need to:

- Develop a list of your services and save it in a .txt file

- Develop a list of the cities that you operate in (your primary city of service and the smaller surrounding towns) and save it in a .txt file
    - Go to www.mergewords.com
    - Paste your list of cities in column 1
    - Paste your list of services in column 2
    - Press the "Merge!" button
    - The tool will generate a list of all your services combined with your cities of service
- Go to Google.com and search "Google Keyword Tool" or go directly to https://adwords.google.com/o/KeywordTool
    - Paste your list of merged keywords into the "word or phrase" box
    - Press "Submit"
- You will now see a list of each of your keywords with a "search volume" number beside it
- Sort the list from greatest to smallest

You now have a list of the most commonly searched keywords in your area.

With this list, you can map out keywords to specific pages on your website and rest assured that you are basing your strategy on opportunity rather than a guesstimate.

Most Commonly Searched Industry Keywords

Mentioned below is the list of the most commonly searched keywords for law firms.

Below you will find the list of the most commonly searched keywords for the dental industry.

Most Searched Law Keywords:

Lawyer	1,65,000
Attorney	1,35,000
Immigration lawyer	74,000
Lawyers near me	90,500
Personal injury lawyer	60,500
Car accident lawyer	49,500
Divorce lawyer	55,000
Law firms	33,100
Family lawyer	32,500
Real estate law attorney	40,500
Personal injury attorney	27,100
Business lawyer	27,500
Tax attorney	12,100
Auto accident lawyer	12,500

Based on this data, in order to get the most from the Internet from an SEO perspective, you will want to create content on your website for the following keyword combinations:

Your City + Lawyer
Your City + attorney
Your City + Immigration lawyer

Your City + Lawyers
Your City + Personal injury lawyer
Your City + Car accident lawyer
Your City + Divorce lawyer
Your City + Law firms
Your City + Family lawyer
Your City + Business lawyer
Your City + emergency lawyer
Your City + Real estate law attorney
Your City + Personal injury attorney
Your City + Tax attorney
Your City + Auto accident lawyer

Now that you have mapped out the pages to be included on your website, you can start thinking about how to optimize each of those pages for the major search engines (Google, Yahoo and Bing).

How to Optimize for Ranking in the Organic Listings

Step 1 – Build the website and obtain more placeholders on the major search engines.

A typical law firm website has only 5-6 pages (Home – About Us – Our Services – Coupons – Testimonials - Contact Us).

That does not create a lot of indexation or placeholders on the major search engines. Most service providers, for example, provide a wide variety of services, as covered in the Keyword Research section of this chapter.

By building out the website and creating separate pages highlighting each of these services offered (combined with city modifiers), a law firm can get listed on the search engines for each of those different keyword combinations.

Here is an example:

- Home – About – Coupons – Contact Us
- Sub-pages for each service – Miami Divorce, Miami Child Support, Miami Child Custody, Miami LGBT Divorce, Miami Divorce Filing, ETC

Law firms often provide services in a large number of locations outside of their primary city. In order to be found on the major search engines for EACH of those sub-cities, additional pages need to be created:

Sub-pages for each sub-city serviced – Family Law, Child Custody, Asset Division, Mediation Services etc.

Step 2 – Optimize Pages for Search Engines:

Once the pages and sub-pages are built for each of your core services, each page needs to be optimized from an SEO perspective in order to make the search engines understand what the page is about.

Here are some of the most important items that need to be taken care of for on-page search engine optimization:

- Unique Title Tag on each page
- H1 Tag restating that Title Tag on each page
- Images named with primary keywords

- URL containing page keyword
- Anchor Text on each page and built into Footer – Miami Divorce Lawyer
- XML Sitemap should be created and submitted to Google Webmaster Tools and Bing Webmaster Tools

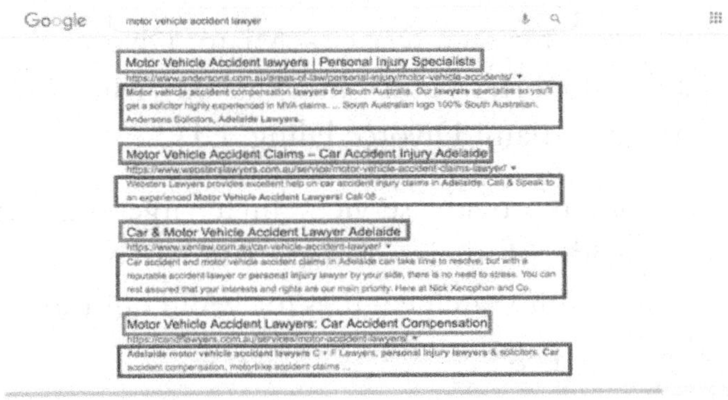

How to Build Up Website Authority

Once the pages are built and the "on-page" SEO is complete, the next step is getting inbound links so you can rank on page one for your most important keywords.

Everything we have discussed to this point is sort of like laying the groundwork. The pages need to be in order to even be in the running. However, it is the number of QUALITY inbound links and web references to those pages that is going to determine placement.

30% of SEO is On-Page type work
The other 70% is Link Building

Building the pages is just the beginning. The only way to

get your site to rank above your competition is by having MORE quality inbound links and citations to your site.

He Who Has The MOST Quality Inbound Links WINS!

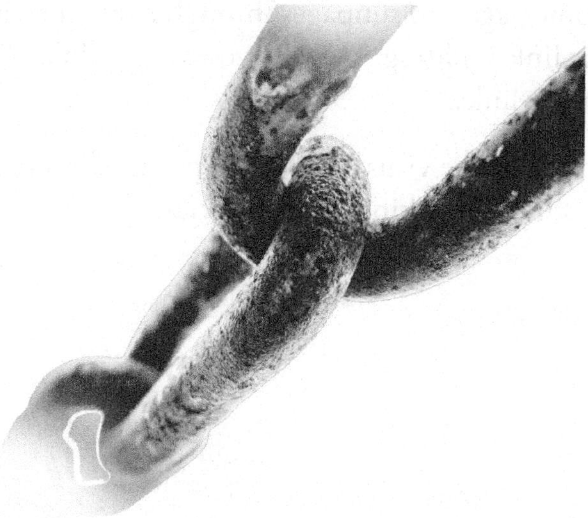

Again, if there is any secret sauce to ranking well in the search engines, it really is links and authority. The major caveat? You can't just use garbage links. You don't want to just have a thousand links. When I say links, I'm referring to other websites hyperlinking to your website, which I'll explain a little bit more with specific examples.

The latest algorithm changes (Google Panda and Google Penguin) involve Google trying to prevent spam. A lot of Internet marketers and SEO coordinators realize it's all about the links. That is what the Google algorithm was built upon. They figured out ways to get a variety of links with random anchor text pointed back to the pages that they want to have ranked. Google has recognized that if those links are not relevant, then they don't add any value to the Internet.

Bad or irrelevant links can actually hurt your ranking more than help it. It's about getting quality, relevant links back to your home page and subpages through content creation and strategic link-building. How do you get the links? Where do you get the links?

Take a look at the visual below as a point of reference. I call this my circle of linking opportunities:

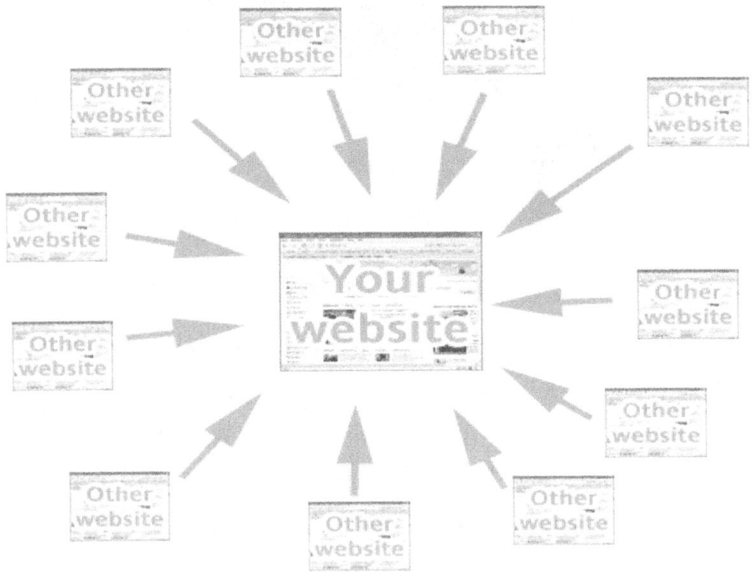

1. **Association Links** – Be sure that you have a link to your site from any industry associations that you belong to (Ex. Business associations, Chamber of Commerce, Networking Groups, etc.).

2. **Directory Listings** – Get your site listed on as many directory type websites as possible (Angie's List, Yahoo

Local Directory, Judy's Book, Yelp.com, Findlaw, Avvo, Justia etc.)

3. **Create Interesting Content/Articles** – This is probably the #1 source of inbound links. For example, you can write an article about a particular service offering in your specialty and push it out to thousands of people through article directory sites that may each contain a link back to a specific page on your site.
4. **Competitive Link Acquisition** – This is the process of using tools like Raven Tools, SEO Book and others to see what links your top competitors have, and then get those same or similar links pointed back to your website.

Directory Links - There's a number of what I like to call "low-hanging fruit" links.

It all starts with your online directory listings.

Some examples include Google Maps, Yahoo Local, City Search, Yelp.com, Judy's Book, Best of the Web, Yellow Pages, Hot Frog, Findlaw, Avvo, Justia, and the list goes on. All of those online listings let you display your law firm name, address, phone number and a link back to your website. Some of them even allow reviews.

For the most part, adding your law firm information to those directories is completely free of charge. You want to make sure that you have your company listed on as many of the online directory listings as possible for authoritative linking reasons.

They're also valuable from the Google Maps optimization

perspective because they give you citations, which are very important for getting ranked on the Google map.

A great way to find additional online directories to add your company to would be to run a search in Google for "Your Company Type – Business Directory" or "Your City – Business Directory". This will give you a great list of potential directory sites to add your law firm to.

There are also tools for this like BrightLocal or White Spark that can provide you with a list of directory sources based on the legal industry. After beginning with online directory listings, you want to look at any associations you're involved with.

Association Links – In the visual, I reference some of the business associations. I'm assuming you are involved in some type of association, whether it is the national industry association, the local chapter or some other group affiliation.

Visit the websites of those organizations and get listed in the member section. This will give you citations and the opportunity to link back to your website.

Non-Competitive Affiliated Industries and Local Businesses - You can work with colleagues that have affiliated industry type businesses.

Being in legal, go to the law firm Association in your area and ask if they will post a link to your website on their own site and vice versa. Utilizing your resources and teaming up with relevant companies will add more authority to your domain.

Supplier Sites - Look at the suppliers or vendors you purchase from and try to coordinate a deal with them. Oftentimes, the places where you buy your merchandise or services will have a section on their website that mentions their value add resellers. You can get a link from those.

Social Media Profile Links - The other "low-hanging fruit" links are social media profiles. We have a whole chapter about the power of social media and how you can harness it to get repeat and referral business.

Simply from a link-building perspective, you should set up a Facebook page, Twitter account, LinkedIn profile, Google Plus page, Pinterest profile and a YouTube channel and place a link to your website on each.

All of them will allow you to enter your law firm's name, address, phone number, a description and, of course, a place to put your website address.

Local Association - Other local associations that you're involved in. If you're a member of the Chamber of Commerce, a networking group like BNI (Business Networking International), or if you're involved with a local charity, find out if they list their members on their websites. Another great place to get links is by typing in your city directory.

Competitive Link Acquisition - You might be surprised that if you really tackle these elements and you don't do any of the other things we have discussed, you will notice that you've probably got enough links to outrank your competition in your area.

I want to share some additional thoughts and strategies on how you can accomplish even more from a link building perspective. A very powerful strategy that you can implement is called Competitive Link Acquisition.

The way I like to think of it is that if quantity inbound links are the secret sauce to outranking your competition, and if we could figure out who's linking to your competition or what links your competition have, and we can get those same or similar links pointed back to your website, then you can outrank them, because you'll at that point have more authority.

Competitive link acquisition is the process of figuring out who is in the top position for your most important keywords, reverse engineering their link profile to see what links they have, and getting those same or similar links pointed back to your website. A simple way to do this is just to go to Google.com and type in "your city + your service," and find out who is in the top few positions.

Let's take a look at the number one placeholder. He's there because his website is optimized well and Google knows he should be ranked well based on the quality and quantity inbound links compared to the competition.

Once you know who he is, you can use a couple of different tools such as Raven Tools, Majestic SEO, Back Link Watch, etc., and you can take their URL, input it into your tool of choice, run the report, and get a list of links in return.

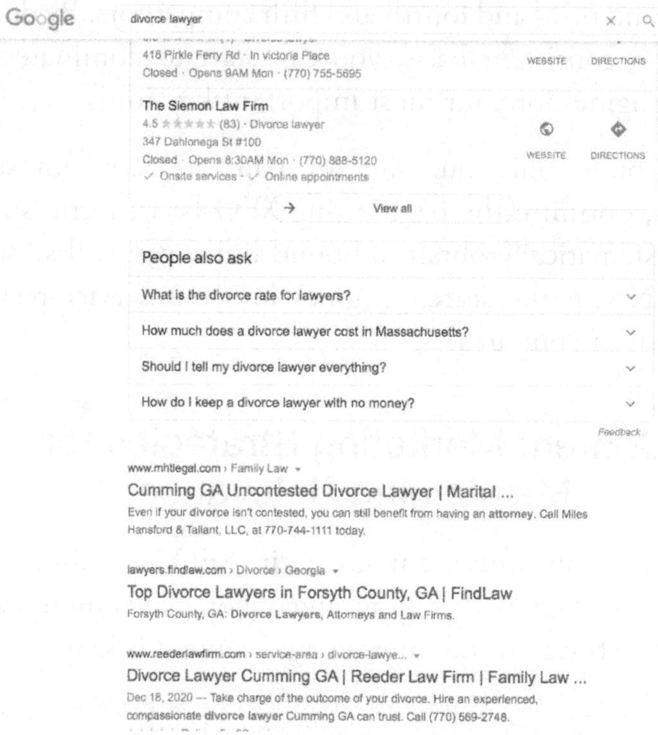

So, your number one competitor is competitor.com. Google spits out a list showing they have 392 inbound links.

- He's got a link from the local Chamber of Commerce
- He's got a link from the PHCC
- He's got a link from an article that he posted in the local newspaper
- He's got a link from the local networking chapter

By analyzing the types of links he has, you can systematically mimic those links and get them pointed back to your website.

Don't just do this for your first competitor, but also for your second and third and fourth and fifth competitors. By doing this on a consistent basis, you can start to dominate the search engines for your most important keywords.

If you build out your site for your services and sub-services, optimize the pages using SEO best practices and then systematically obtain inbound links, you will start to DOMINATE the search engines for your service-related keywords in your area.

Content Marketing Strategies for Maintaining Relevance

Another highly important factor in SEO is maintaining relevance in your market by adding ongoing relevant updates to your website. In the Internet age, content is king.

Google Loves Fresh Content!

In some cases, with the changes in the algorithm, just because you've got a great website with the right title tags and all the best links, you may get discounted if they're not seeing fresh information posted on a consistent basis.

Google loves fresh content, and it is important to have a methodology where you are creating and posting content to your website on a regular basis. I want to give you a framework for figuring out what kind of content you could write, why you should create content, and how you can do it consistently.

First, you need to understand and accept that you need to become a subject matter expert. You might not consider yourself a writer or a content creator, but you are a subject matter expert.

There are things you know that the general population does not. You're an expert when it comes to the service you are offering, and you have a team of people who are proficient in this area as well. You can create content on the topic that you know most about.

You can write about a variety of different topics including 1-2 relevant keywords with the same. You may not think it at first, but there are a lot of different topics in your specialty you can come up with to create content about.

Types of Content

You should also consider that content doesn't have to be just written words. It's doesn't have to be just articles. Content can come in a variety of forms. The most popular are going to be articles, photos, videos and audio files. Stop and think about what content creation method works best for you.

Some people are great writers and that's their strength. Other people like to be on camera. I personally like to create videos. I'm very comfortable creating videos. Other people can talk (and they can talk your ear off about whatever topic they are passionate about!)

You can create content in many different ways. Because it is what I enjoy, I'll use video as an example. A business director can set up a camera and record himself explaining the services his company has to offer in the market, in the same manner that he would explain it to a customer.

Now you'll actually have multiple pieces of content. You'll have a video, which can be uploaded to YouTube, Vimeo, Meta Café, etc. This one piece of content can create multiple invaluable links to your website.

You can also take that video, save the audio portion of it, and you've got an audio clip. You can upload that audio file to your website and post on other various sites.

You can use a transcription service like Castingwords.com, for instance, where you upload the audio or video file and somebody converts it to text.

For a couple of bucks, you'll have a complete article comprised of what you said. Now you've got a piece of content you can post to your blog. You can put it on eHow or one of those other article directory sites.

Content Consistency

You want to create content on a consistent basis, using the blog on your website as the hub to post it, but then syndicating it to various sources.

Syndicating it to article directory sites if it's in text form, and sending it to video sites like Vimeo, Metacafe and YouTube.com if it's in video form. Doing this keeps the content fresh on your website/domain and creates a lot of authority, which is really going to help with the overall ranking of the website on the search engines.

You want to make sure you're appropriating each one of these link-building opportunities to maximize your rank-potential in your area. You might be surprised that the services you offer are highly competitive from a SEO perspective. There are a lot of law firms who want to rank for the same keywords, and many of them have invested heavily in the Internet and in getting themselves higher in the search engines.

Now that you've built out your website, you've optimized it correctly, and you've got an ongoing link-building and content development strategy in place, you want to start looking at Google Maps Optimization and getting ranked on the Google Map.

SIX

Google Maps Optimization-

How to get ranked on the Google Map in your area

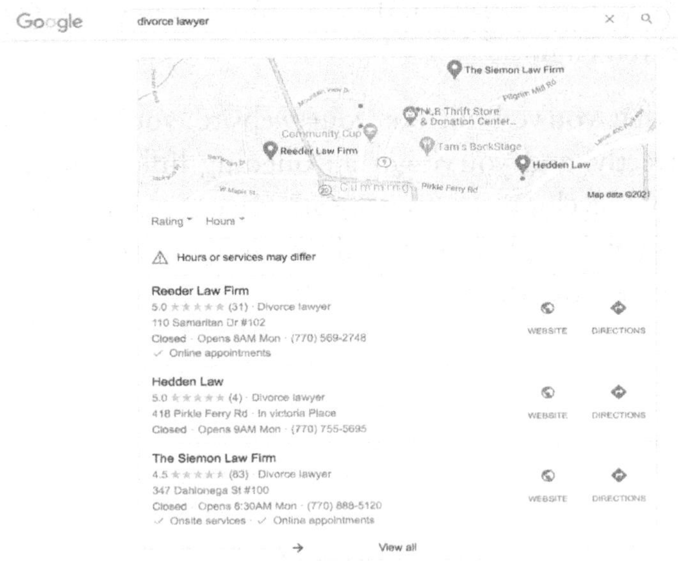

The Fundamentals of Google Maps Ranking (NAP, Citations, Consistency and Reviews)

Getting listed on the first page of the Google Map for "Your City + Service" comes down to four primary factors:

- Having a claimed and verified Google Map Listing
- Having an optimized Google My Business listing for the area you operate in
- Having a consistent N.A.P. (Name, Address, Phone Number Profile) across the web so that Google feels confident that you are a legitimate organization located in the place you have listed and serving the market you claim to serve.
- Having reviews from your clients in your service area

If you have each of these four factors working in your favor you will SIGNIFICANTLY improve the probability of ranking on page one of Google Maps in your market.

How to establish a strong Name, Address, Phone Number profile

As I mentioned above, having a consistent Name, Address, Phone Number Profile across the web is essential for ranking well on the Google Map in your area. Google sees it as a signal of authority.

Rather than jumping directly into claiming your Google Map listing and citation-building, it's critical that you start by determining your true N.A.P. so that you can ensure that it is referenced consistently across the web.

When I say making sure it's consistent, you want to be certain that you are always referencing the legitimate name for your law firm.

If your company's name is "Bob's law firm", you must always list it as "Bob's law firm," as opposed to just "Bob's Legal Services."

The other thing you should be aware of is that there is a lot of misinformation about how to list your law firm name online. You may read information suggesting that you keyword your name.

For example, if your name is "Bob's law firm," somebody might tell you it would be really smart if you just added to the title of your company "Bob's law firm | Dallas Lawyers," for instance.

While that may have worked back in the day, it's no longer an effective strategy. It's actually a violation of Google Places' policies and procedures.

Make sure you list your exact law firm name the same way across the board on all your directory sources. Also make sure that you use the same phone number in all those places. I'm a big advocate for tracking phone numbers and what is happening with your marketing. But, when it comes to your online directory listings, you want to use your primary law firm phone number that you've been using from the beginning.

Don't try to create some unique number for each one of your directories. What that does is confuse your name/address profile. It will hurt you.

Use your primary phone number in all those places, use your exact law firm name, and use your principal address, written the same way. If your business is located at "1367 South West 87th Street, Suite Number 105," make sure you list it just like that every single time.

Don't Forget the Little Details!

Don't neglect to include the suite in one place and then put it on in another. Don't spell out "South West" in one place and put "SW" in the other. We are driving for a consistent name/address profile across the web.

A good way to figure out what Google considers to be your N.A.P. is to run a search on Google for "Your law firm" and see what is being referenced on the Google Map.

See how that compares to the other high authority sites like YP.com, Yelp.com, Angie's List and others. Look for the predominant combination of N.A.P. and reference that for all your directory work going forward.

How to Properly Claim and Optimize Your Google My Business Listing

Below you will find a step-by-step guide for checking, claiming and managing your Local Business Listings on Google.

1. Go to https://www.google.com/business/

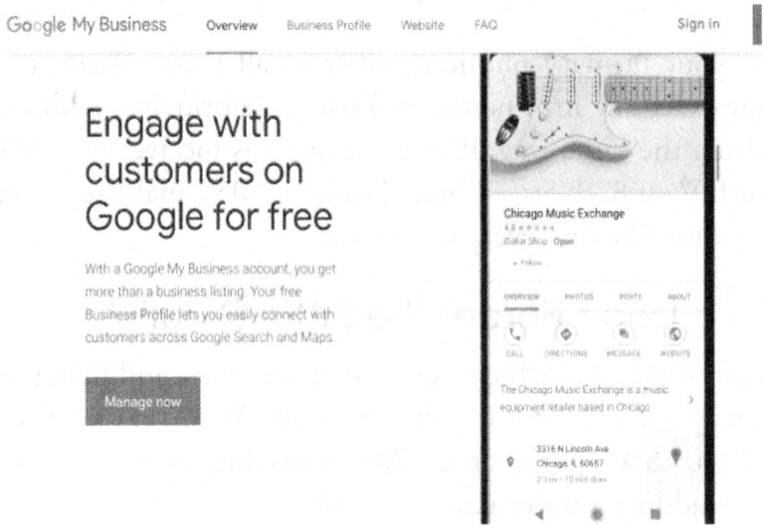

2. Create an Account and claim your business

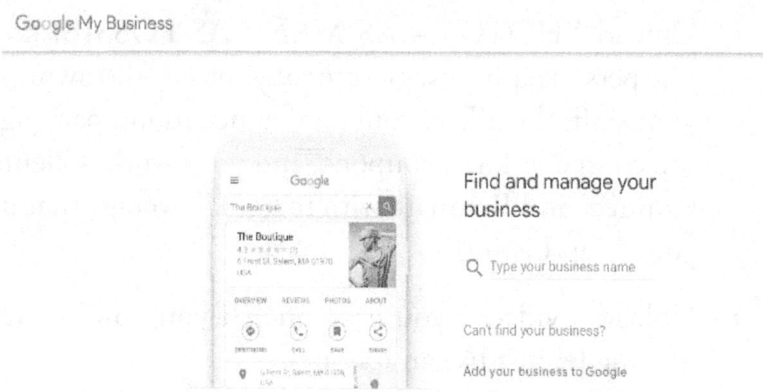

3. Enter your law firm address and all pertinent information

4. Choose a verification method

- By postcard
- By phone
- By email
- Instant verification
- Bulk verification

5. Once you've created a profile, go to your Google My Business Dashboard and fill in all the necessary information to optimize your profile.

- Update Your law firm Name to Read "Company Name") – e.g. Bob's law firm. Don't add any additional keywords here

- Add your Website Address – This will create an important inbound link
- Upload PHOTOS – AS MANY AS POSSIBLE – Use personal photos, pictures of yourself (the owner), your staff, the office, your conference room, parking, the company logo, coupons, and your work. Clients Connect and Resonate with images. Leverage that in your Map Listing
- Upload a video if you have one (If you don't – Get one made! Just use an iphone)
- List your hours of operation and services offered

Optimize your Google My Business Listing

You'll manage your law firm business listing from your Google My Business Dashboard.

Here's where you'll make changes to your law firm information, gain insights into how popular your law firm is by seeing how many times your profile has been viewed on Google Maps.

There are a number of Best Practices you need to be aware of to properly optimize your Map Listing

1. Company Name – Always use your legal Company Name – don't cram additional words into the name field. Ex. If your company name is "Joe's Car Repair," don't try to

put additional keywords like "Joe's Car Repair – Dallas". This would be against the Google Places guidelines and will reduce your probability of ranking.

2. Address – On the "Address Field" use your EXACT legal address. You want to ensure that you have the same address listed on your Google Places listing as it is on all the other online directory listings like YellowPages.com, CitySearch.com, Yelp.com, etc. The consistency of your N.A.P. (Name, Address, Phone Number Profile) is very important for placement.

3. Phone Number – Use a local number (not an 800 number), and make sure it is your real law firm number rather than a tracking number. We find that 800 numbers don't rank well. If you use a tracking number, it won't be consistent with your other online directory listings and will result in poor ranking unless you use a local area code and consistently use it everywhere.

4. Categories – You can use up to five categories, so use ALL five. Be sure to use categories that describe what your business "is" rather than what it "does". So, you can use "Lawyer" "Family Law" etc. rather than "Paralegal Services". The latter would be considered a violation of Google's regulations and would hurt rather than help you.

5. Service Area and location settings – Google offers two options here:

    - No, all clients come to my location
    - Yes, I serve clients at their location

6. For Example: If you owned a home services business, you need to select "Yes, I serve…" because clearly you and your technicians are visiting the customers at their location. Not doing so can result in a penalty on your listing. On the other hand, if you have a business shop, it is most likely that customers come to your location for the goods/services provided and, in this case, you need to choose "No, all customers...".

7. The next option is "Do not show my address". If you work from a home office, it is required that you select "Do not show my address." Not doing so puts you at risk of having your listing deleted.

8. If you don't have a business address or a home address to list, the only other option is a virtual office. Unfortunately, P.O. Box addresses and mailboxes don't tend to rank well.

9. Picture and Video Settings – You can upload up to ten pictures and five videos. Use this opportunity to upload authentic content about your law firm. It's always best to use real photos of your team, office, and facility rather than stock photos.

10. Pictures – You can get more juice from this section by saving the images to your hard drive with a naming convention like "your city + Family Law – your company name," rather than the standard file name. You can also create geo context for the photos by uploading them to a video sharing site like Panoramio.com (a Google Property) that enables you to Geo Tag your photos to your law firm's location.

11. Videos – Upload VIDEOS. They don't have to be professionally produced and will resonate well with your customers. A best practice is to upload the videos to YouTube and then Geo Tag them using the advanced settings.

Once you have optimized your listing using the best practices referenced above, you want to be sure that you don't have any duplicate listings on Google Maps.

Duplicate Listings

We have found that even just one or two duplicates can prevent your listing from ranking on page one. In order to identify and merge duplicate listings, run a search on Google for "Company Name, City".

To clean up duplicates, click on the listing in question and then click "edit business details."

Click "This is a duplicate" to let Google know that the listing should be merged with your primary listing.

If you follow these best practices, you will have a well-optimized Google Maps listing for your Business.

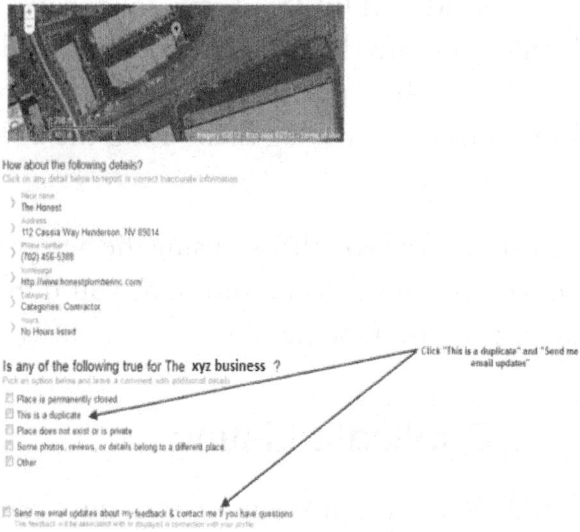

How to Develop Authority for your Map Listing via Citation Development

Now that you have claimed your Google My Business Listing and optimized it to its fullest, you need to build authority.

Having a well-claimed and optimized local listing doesn't automatically rank you on page one. Google wants to list the most legitimate and qualified providers first. So, how do they figure out who gets the page one listings?

Well, there are a number of determining factors, but one of them is how widely the company is referenced on various online directory sites such as Yellow Pages, City Search, Yelp and others.

Citations are web references to your law firm name, address and phone number. You can add citations in a variety of ways. There are directory listings that you should claim manually and others that you can submit to via submission services like Universal Business Listing or Yext.com.

My personal preference is to claim listings manually, ensuring that I am in control and can make updates/edits as needed.

TOP Citation Sources to Claim Manually:

- Google My Business
- Bing Local
- Yahoo Local
- City Search
- FourSquare.com
- Yelp
- YP.com
- Merchant Circle
- Manta
- Findlaw
- Justia
- Avvo

List of the Top Citation Sources for Businesses

1) Google

Google is probably the most important and most talked about place to list your local law firm. Getting citations from many of the sites below (as well as ratings) can help boost your business' listing in Google.

2) Yelp

Yelp is one of the most popular social networking, directory and review site. Aside from counting as a citation for your business in the eyes of major search engines, this site can deliver quite a bit of traffic on its own. However, law firm owners using Yelp will need to learn to deal with the occasional nasty review. We also do not recommend paying for any of the upgraded services.

3) Foursquare

A popular way to check-in to various locations using a smart phone. This can also provide a valuable citation for your local law firm.

4) Universal Business Listing

A local listing service – UBL.org (along with Localeze below) is one of the major players in the effort to only fill in your information once while getting listed on multiple yellow pages sites, directories and social networking/review sites. It saves time and effort but may be slower than going directly with the individual sites (see Localeze and InfoUSA below).

5) Yahoo Local

Yahoo's local directory tied to Yahoo Maps.

6) Local.com

Business listings, event listings, coupons and reviews.

7) CitySearch

One of the most authoritative local directories.

8) Bing Local

Bing's local business listing service integrated with maps of cities and towns.

9) Craigslist

Some recommend creating classifieds for your business on popular sites such as Craigslist. There's some disagreement over whether this is effective from an SEO point of view.

10) GetListed.org

Convenient way to identify where you are and are not listed in major directories. Provides referrals to Universal Business Listings and Localeze as well as consultants (if you need extra help).

11) DMOZ (Open Directory Project)

A free and authoritative index (in the eyes of Google) that is managed by volunteers.

If you can get your business listed, this helps with an authoritative backlink (but not necessarily a local citation). It can be difficult to get a new listing due to the limited resources and large volumes of submissions.

12) Superpages

One of the many Internet Yellow Pages directories (IYP). Includes business listings, people search, reviews and local deals.

13) Localeze

A multiple local listing service.

14) InfoUSA

A multiple local listings service.

15) Your local Chamber of Commerce

Joining your local chamber of commerce can often get you a business listing (and a citation for local SEO purposes)

16) InsiderPages

Local directory and rating site.

17) Merchant Circle

Local directory and rating site.

18) Best of the Web

A popular directory with free and paid listing options – specifically for local, they have a Best of the Web Local directory.

19) Yellowpages.com

Internet yellow pages (also YP.com).

20) Judys Book

Social search and online yellow pages

21) Business.com

Business.com provides business information but also has a business directory.

22) Better Business Bureau

Your local Better Business Bureaus will usually charge for membership and provide a link to your business.

23) DexKnows

Business and people directory.

24) Your local newspaper's website

Getting an article, business listing or classified ad optimized with your local information and a link can provide a citation for your business.

25) Yellowbook.com

Internet yellow pages.

26) HotFrog

A business directory with free and fairly inexpensive paid listing options.

27) Crunchbase

A listing of technology companies that is user generated.

28) Angie's List

Service provider directory.

29) Jigsaw

Businesspeople and company directory.

30) iBegin

US and Canadian business directory.

31) Wikimapia

Wiki-based directory of places including schools, businesses, and more – laid out on maps.

32) CitySquares

Local business directory with ratings.

33) InfoSpace

Business and people listings.

34) MagicYellow

A straightforward Internet yellow pages directory.

35) Whitepages.com

People and business listings.

36) Manta

Company profiles.

37) EZLocal

Local business listings and ratings.

38) BrownBook

Local business listings and ratings.

39) CityVoter

Vote for favorite businesses.

40) ShopCity

Local business listings.

41) YellowBot

Local listings and ratings.

42) MojoPages

Social networking and review site (like Yelp).

43) Tupalo.com

International social networking and review site.

44) GetFave

Business directory, including featured listings (with additional content such as videos and pictures).

45) BizJournals

Business journal that includes business directories for certain US cities.

46) Tjoos

Online store listings and coupons.

47) JoeAnt

Website directory.

48) Zidster

Products, services or business listings.

49) TrueLocal

Business directory – seems to have sparse listings.

50) ZipLeaf

Network of international business directories.

51) WCities

Places and events for cities and towns, including ratings.

52) Naymz

Personal branding site.

53) ZoomInfo

Database of people and companies.

54) Yellowikis

Wiki-based business directory.

55) GoMyLocal

Yellow pages/local directory.

56) Fast Pitch Network

Online business networking site.

57) City Slick

Free business directory.

58) yellowpages.lycos.com

A general directory.

59) Home Advisor

Directory of service companies (includes a "seal of approval").

60) Findlaw, Justia, & Avvo

Legal specific directories where you can list for free.

By securing these high-quality citations you will boost your authority and highly improve your probability of ranking in the Google Map Listings. The next critical step is to get online reviews!

How to Get Online Reviews: Real Reviews from your Real Customers in your True Service Area

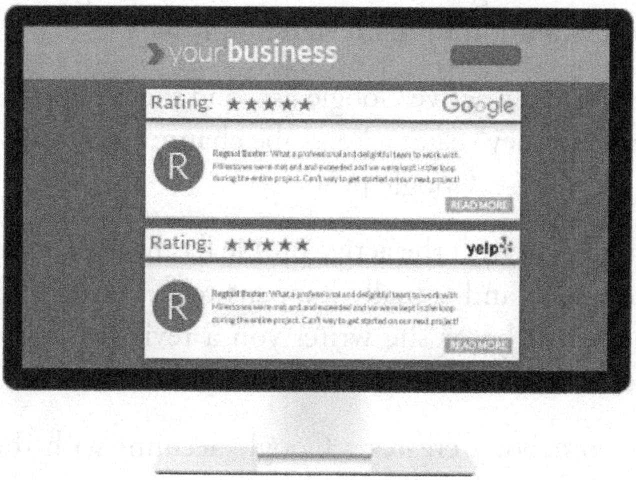

The next critical component for getting ranked on the Google Maps, after you've claimed and optimized your listing, you've established your N.A.P. and you've developed your citations across the web, is obtaining reviews. You need to have real reviews from your real clients in your true service area.

Keep it Real

First, I want to point out that you shouldn't fill the system with fake or fraudulent reviews. You do not want to create bogus accounts and post reviews to Google Map, Yelp, City Search, etc. just for the sake of saying you've got reviews. That's not going to help you. You need real reviews from your actual clients in your true service area.

You might be thinking, "Well, how is that important?" or "How would Google know the difference?" Google is paying very close attention to the reviewer's profile.

If somebody is an active Google user and they've got a Gmail account, and they've got a YouTube channel, typically that's all connected to a Google profile.

Say that person with the active profile has had their account for seven years and actually happens to be located in your service area. If he or she writes you a review, it would be considered credible and will count in your favor.

Now, if somebody creates a Google account with the sole intent of writing a review, it obviously is not credible, and Google is capable of catching on to that. That account has no history associated with it and it was originated right at your office IP address. That review is going to be flagged as a bogus submission.

It is important to have an authentic strategy where you are connecting with real people who will write your reviews. You don't want to try and play the system. Google is fully aware, and so is Yelp and a number of other popular online review sites.

Getting Reviews

With that said, how can you get reviews? What kind of process will you need to actually get reviews from your real clients in your real service area? Here's the strategy that we advocate

First of all, have some review cards printed up (a sample is referenced later in this chapter). It's basically just a simple document with your company logo, and a short and sweet thank you note.

"Thanks so much for your business. We appreciate the opportunity to serve you. We'd love it if you would write us a review." Then give them a link to a page on your website where they can write you a review.

You will want to do some homework on the front end. Be sure you have a page on your website that is clearly meant for reviews: yourcompany.com/reviews. On that page you'll have links to the various places where people can write your reviews.

You'll want to have a link to your Google map listing, Yahoo local listing, Angie's List listing, City Search listing and any others that you may have. The reason you want to really have a variety of places where people can write those reviews is twofold.

Yes, you want to have a lot of reviews on the Google map. But Google is also looking at the reviews that you have on other websites like Yelp and Angie's List. They're looking at the reviews that you have on Yellow Pages and other pages.

You need to diversify where you're getting reviews from your customers. It looks more authentic to have 12 on Google and 17 on YP.com, than it does if you just have 72 reviews on the Google map.

You want to make it easy and you want to give people options.

Make it Easy

The other thing you want to bear in mind is that different people use different systems.

I am personally a big Google user. If you sent me an email or gave me a card that said, "Please write me a review" and provide me with various options, I'm going to say, "OK,

Google." Click Google. Write my review.

Some people, however, don't have Google accounts. They're not active Google users, but they may be heavily involved in Angie's List or big-time reviewers on Yelp.

They're going to have active accounts somewhere. It would be much easier for them to write the reviews where they already have an existing account. The easier and more convenient you make it for people, the better. It's going to bode well in your favor.

Like we mentioned, Google is looking at the reviewer profile. If you only give them one option, and that's the Google map, but they happen to be a Yelp user without a Google account, they will have to go out of their way to create an account to write the review.

This is not likely to happen. But, let's say they did decide to create an account. That review is not going to count for much because there's no active profile.

By providing options, the Yelp user that has a reputation for writing reviews and decides to write one for you is going to make a difference. That review is going to stick as opposed to being filtered. Make it easy for them to choose the one that's going to be easiest for them.

Now Let's Get Back to the Strategy

Phase one, print out review cards. Have your team hand them out after a meeting or closed case. "Hey, thanks for your business. I just want to leave this with you. If you'd be willing to write us a review and share your experience, we would really appreciate it."

It's great. You're showing appreciation. You're holding yourself accountable because you're asking for feedback. By doing that on a consistent basis, you are likely to catch some fish.

The next thing you'll want to do, just to get a nice little bump in the number of reviews that you have, is to develop an email list of your circle of influence.

Your circle of influence is going to be your most recent clients, the clients that have been using your services for quite some time, your family members, and your friends. People that you know, like, and trust, who would be willing to act on your behalf.

Put together that email list in an Excel sheet. It might be ten contacts, or it might be 700 contacts. Include the names and email addresses of these folks. Then, use a tool like Constant Contact or MailChimp or another email marketing tool to send an email blast with the following message:

Email Subject: Thanks for your business!

Name,

I wanted to shoot you a quick email to thank you for your business and let you know how much we appreciate the opportunity to serve you!

Our goal is to provide 100% customer satisfaction and exceed your expectations every step of the way. I certainly hope that we did just that!

If so, it would really help us out if you'd be willing to post a review for us online at one of your favorite online review sites. Below are a few sample direct links where you could write a public review about your experience with us

- https://support.google.com/maps/answer/6230175?co=GENIE.Platform%3DDesktop&hl=en
- Yelp - https://www.yelp.com/writeareview

Thank you again! We really appreciate your support!

Best Regards,

John Smith

Again, save them the time of having to find the websites on their own by providing some links to the various places to where they can write reviews.

By sending this email, you're going to create a little bump in your online review profiles. Again, reviews are important. Getting ten reviews on Google Maps is essential.

It makes a huge difference in how you rank, and it gives you a different perception in the mind of your consumers. You want to get to past that ten review threshold almost immediately.

Doing that helps you get real reviews from real people that have real online profiles. Again, you want to have a systematic process in place where you are asking for reviews on a consistent basis from the customers that you are serving on a daily basis. The best way to do that is to request an email address from your customers, either at point of service or after service.

Establishing your Email Database

We have found that the best time to ask for that email address is at the point of booking the consult. If you wait until after the consultation is rendered your team would be saying "OK, thanks for the opportunity to review your case, by the way could you please give me your email address". They are going to say, "Why do you need my email address?" "Oh, because I want to ask you for a review or..." There is a lot of resistance to it at that point in the sales funnel.

However, if you move into the front where somebody calls in and says, "Hey I need to schedule a consult I was in an auto accident" you can respond, "We can get something scheduled right away. Let me gather your information."

This is the perfect time to get the email address. Typically, you get their name, address, and the phone number. Well, you can just add one more step at that point and request an email address as well. You can tell them that it is so you can send a confirmation.

That's how you start to develop a database of emails. We are going to talk about email marketing later in the book as part of your online marketing plan, but for this purpose, you need an email address so that you can send a message after service thanking them for their business and asking them to write you a review.

The number of reviews that you have from actual clients is going to increase exponentially if you repeat this process regularly. This is how you are going to start to really dominate the Google Map, because reviews and citations work in harmony for ranking.

Sample Review Card

Sample Review Request Email

Name,

I wanted to shoot you a quick email to thank you for your business and let you know how much we appreciate the opportunity to serve you!

Our goal is to provide 100% client satisfaction and exceed your expectations every step of the way. I certainly hope that we did just that!

If so, it would really help us out if you'd be willing to post a review for us online at one of your favorite online review sites. Below are a few direct links where you could write a public review about your experience with us:

Google - https://support.google.com/maps/answer/6230175?co=GENIE.Platform%3DDesktop&hl=en

Yelp - https://www.yelp.com/writeareview

Thank you again! We really appreciate your support!

Best Regards,

John Smith

Sample 'Review Us' Landing Page for your Website

If you follow these steps to properly claim your Google Map listing, develop your authority via citation development, and put a systematic process in place to get real reviews from your real clients in your true service area, you will be well on your way to dominating the Google Map listings in your market.

SEVEN

Website Conversion Fundamentals -

How to ensure that your website converts visitors into leads in the form of calls and web submissions

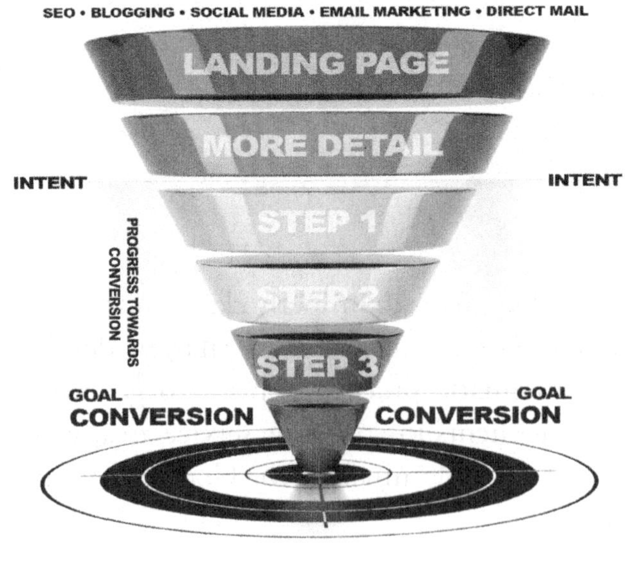

This chapter is all about website conversion fundamentals. We talk about how you need to set up your website, the messaging on your website and the navigational flow of your website to ensure maximum conversion and profitability from your entire online marketing effort.

The way I look at it is, you can have the best Pay Per Click campaign, search engine optimization, and be ranked number one on the Google Map. But, if the content and the structure of your website isn't set up in a way that's compelling for users, then it doesn't give them a reason to choose you over the competition, and it doesn't give them the information that they need to easily say, "You're the law firm that I am going to call for help." It's just not going to do as well as it could.

I want to talk about how we can take the traffic we're going to get from organic and Pay Per Click strategies, and make sure that the website is illustrating the correct message so we can maximize the profitability and revenue of our online marketing strategies.

Conversion Fundamentals

Be real. I talked about how people resonate with real people. They like to see the law firm, the people that they are going to be talking with on the phone and that are going to be going out to their home. So, as often as you can, avoid stock photography. Get a picture of the owner, the team, your office, or the front of the building.

These things really draw people in and it gets them to feel

that they would be working with real people because that is the kind of law firm that people want to deal with.

As for the content of your website, write messaging that draws them in and makes them connect. They're looking for a Personal Injury Firm, Family law firm or other specialy, so when they land on your home page, the first message they see should enforce the fact that they can trust you.

You should write something along the lines of, "Are you looking for a Personal Injury law firm that you can trust? Then you've come to the right place. We're operating on the same principles for the last 30 years: trust, innovation, and excellence."

Connect with them. Give them reasons to choose you and have a call to action (CTA), "Give us a call at this number for immediate service," or, "Click here to take advantage of our online specials and discounts." Remember, they've browsed around the Internet and have seen that there are dozens of law firms that they can choose from.

Give them some compelling information about who you are and why they would want to choose you. Ask them to call now for a consultation, and then draw them into a section where they can get an offer or a special discount. This is going to incentivize them to choose you and make that call right away.

What to Write

When it comes to the copy on the website, you want to address their specific concerns.

For example, if you own a law firm, on the home page, write something generic, "Looking for a trusted law firm?" On the divorce attorney page, sympathize with them. "I know how frustrating it is when you filed a divorce and it is not coming through. You need to make sure that you've got the right assistance."

Write that kind of messaging for each one of the pages on your website including a clear call to action after every block of text saying, "Call now to schedule your appointment," or, "Click here to reference one of our online coupons for a discount on your first service."

Pull them deeper into your website with "About Us" links, special offers, and links to testimonials.

Give them content that makes them think, "This firm knows what they're doing," and draw them deeper and deeper into the website so they're more inclined to take the next step. Tell them why they should choose you over the competition. I talked about this in the "Message Market Media" chapter.

You should also, of course, have a web form on each of the pages of your website or, at a minimum, on the "Contact Us" page.

This is so that if they're not in the modality to pick up a phone, they can simply type in their name, email address, and phone number and let you contact them. Again, make sure that you've got your phone number on the top right-hand corner and that you've got a clear call to action telling them what to do next on every page of your website, under every block of text.

Check out our reviews, download a coupon, look at testimonials.

Explain why they should choose you. Leverage personality. Be authentic. Integrate your photos into your website. It really, really helps with conversion.

Utilize your reviews, testimonials and videos. There's no reason you can't create a simple video for each of the pages on your website, explaining what the service is, and why your law firm can do it best.

Some people are visual, they can see the content on the website, read it and feel fine. Other people are more audible and would prefer to hear the message.

If you can spend the time to provide both text and video, it really helps with conversion. Give them external proof. Take them out to the review sites where they can preview testimonials on Angie's List, Google Maps, etc.

Show them what other people are saying, and you're going to significantly improve your conversion.

Example of a law firm Site that is Built to Convert

Above the Fold Section:

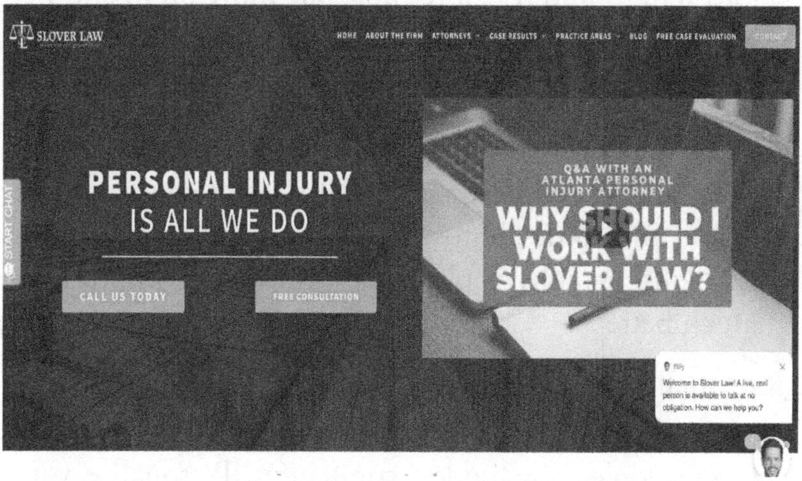

Internet marketing involves a lot of little things that are performed in sequence to get people to call your law firm when they are in need of your service. At the end of the month, it all comes down to the amount of calls you received and how much business was booked, right?

1. Law firm logo should always be in the top left hand side of the page. Their logo here is the perfect size. Sometimes clients tell me they want their logo to be triple this size. The reality is that few searchers know you from your law firm name, so occupying too much space with just your logo is a waste of valuable webpage real estate.

2. Your phone number is VERY IMPORTANT for the credibility of your law firm. It should be as close to the top right hand corner as possible. Make sure it's large and easy to find. Try not to make people search for it. It's frustrating for searchers and you have just a few seconds

for them to find it before they may move on to another website. People always look to the top of the page for that vital piece of info.

3. Professionally shot photos. For a small investment, you could and should have a professional photographer come in and take some photos. You will use them everywhere. DIY photography is ok, but a professional photo is so much better.

4. A small blurb of text confirming the family-owned and operated company really brings it all together. People buy from people, not hidden companies. Personalize your website as much as possible. Your website is a marketing tool and its job is lead capture and to bring down as many buying barriers as possible.

5. Main navigation. Your website's main navigation should be easy to find and the links should be clearly descriptive. Give people the option of moving around your website. One of Google's algorithms is how many pages a person visits and what their visit length was. Guide them down a path without confusing them. In other words, give them all the information they need in as few clicks as possible, but provide them the option of navigating around your site.

6. Some people want a way to contact you without calling. A contact form above the fold (the top half of the page) is great for capturing clients' info. In the case of this website, they can get a lot of form submissions on a monthly basis. Without the form, those clients may not have even contacted them. It's also a great tool for

building a contact list for email marketing down the road.

7. Get to the point right away without going into too much detail. The first paragraph of your text should give you a brief introduction of who you are and what you do. You can go into further detail on your About Us page.

8. You can't see it here, but this area is a slider graphic. This three-window slider is a nice visual effect that adds movement to the page and delivers on three core services or important messages that you want people to know about.

9. Social media icons are a great tool because it allows potential visitors to see another side of your law firm. It's a great place to publish more videos and photos. Also, it's a great place to see how your law firm interacts with its community. From an SEO point of view, it helps build your law firm's social signal, something Google is paying

more attention to. Social media is no longer just sexy "marketing speak", it is a must when it comes to online marketing.

10. Don't forget about going mobile. The mobile web is huge and for the first time ever, mobile searching has passed desktop searching in the local market. It's only going to continue to grow. The important thing with mobile is to make it easy and to get all of the important information front and center. Make sure everything is only a click away and always have your 'call us' button on top.

11. You'll see on this example a video of the founder. This is compelling and allows you to connect with the prospective client. It also keeps them on the website longer which time on a website is a ranking factor for Google.

12. Another great tool is to have a live chat. You can either have your team manage this or it can be outsourced to companies experienced in legal chat management

Website Conversion Factor Analysis

I have summarized the positive and negative points of the www.Globalcreativestudios.comwebsite that can affect the User-Friendly Interface of the website and the website conversions.

Homepage Banners

First, "Above the Fold" should be properly optimized. Above the fold areas are the sections of the webpage that are visible without the need for scrolling. This space is critical in regard to the user engagement and as such the most appealing attributes should be showcased in this section. Some of the most important conversion factors that should be present in the above the fold section is the following:

- Attention grabbing headlines
- Bullet points
- Calls to Action
- Form to capture leads

Please check the following Above the Fold banner for a reference:

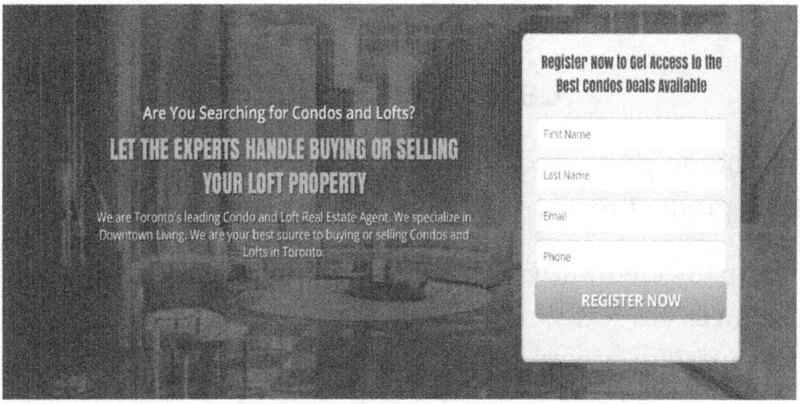

Second, you must include a proper section for "User Engagement" in the "above the fold" section of the website.

Ideally in the header section, there should be a downloadable whitepaper with an interesting title with a "Download Now" button and the user can download the same if they input their name and email address. Please check the following image for a reference:

Law Firm Profile

The home page needs to have a 60% company focus and 40% product focus.

Company credibility is the key to user engagement and any user needs to feel secure about the company before moving forward with browsing through the product section.

Therefore, it is important to have a brief law firm profile on your home page to let users know about "who you are", "why you are different" and "what your business is".

Ideally the home page should contain a brief overview about the law firm and a couple of bullets describing the salient features of the business. Please check the following example for a reference:

Latest News

There should be an enticing Latest News section on your Home Page in a professional press release format as Search engines give preference to sites that have press releases rather than articles on their home page.

The Press Release section should have at least two latest news, with a couple of lines about the news and a visible Read More button. Please check the following reference.

What's New?

» Photo Research Releases Spectrascan® Spectroradiometer PR-788

» Photo Research Announces Merger with JADAK, LLC

» Photo Research Announces Service Partnership with KLight Laboratories

» Tru-Image™ TRU8 High Resolution Imaging Colorimeter Solution

LATEST News

383 Children Cured of Clubfoot since January!

BASE Welcomes Mark Straka, Vice President, to BASE Engineering USA Corp

BASE Wins WorkSafe NB Award for Safety Practices & Leadership

A Beginner's Guide to Propane and Radio Remote Controls

READ More

Blog

A blog is useful as it gives the visitors useful information and also gives credibility to the law firm as the visitor perceives that the business is up to date with the happenings in the industry.

Also in regards to the recent algorithm changes, updating the home page content at regular intervals is critically important in regards to getting higher rankings in SERP's.

This is the reason the website should have a blog section in the home page similar to the following and the blogs should be updated regularly. Please check the following reference:

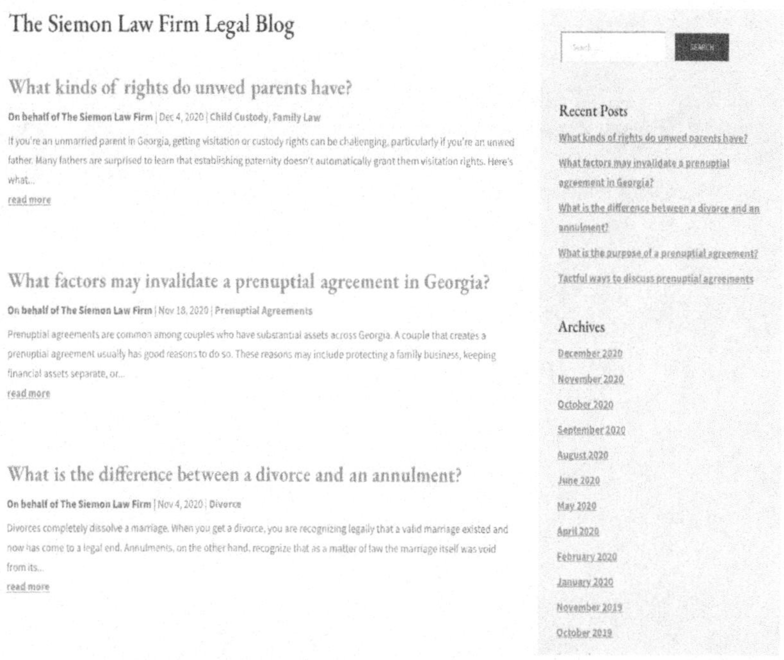

Testimonials

Testimonials are important since they earn credibility from your users.

The testimonials in your home page should have a "Read More" which will redirect to the testimonial page. Please check the following reference:

EIGHT

Mobile Optimization -

How to optimize your website for mobile visitors

More and more of your clients are searching for law firms via mobile devices. Here are just a few eye-opening mobile stats you should be aware of:

- Up to 70% of web traffic happens on a mobile device (CIODive, 2018)
- 80% of users used a mobile device to search the internet in 2019 (Google)
- By 2020, the number of smartphone users is projected to reach 2.87 billion. (Google)
- World Advertising Research Center predicts that by 2025, three-quarters (¾) of users will access the Internet using a mobile device.
- More than one-quarter of all emails are opened on smartphones. (Source: Smart Insights, 2019)
- 40% of mobile users search for local businesses and often include the phrase "near me". (Source: Aumcore, 2019)
- It is predicted that mobile ad spend will reach almost 250 billion dollars by 2020. (Source: Statista)
- 61% of users will have a more positive opinion about your business if you have a good responsive website. (Source: Smart Insights, 2019)
- 57% of users say they wouldn't recommend a company if their mobile website design is bad. (Google)
- By 2021, half of all eCommerce sales will be mobile sales. (Source: 99firms)
- According to BrizFeel, 49% of users use their mobile devices for shopping. (Google)

- 50% of B2B search queries today are made on smartphones and will grow to 70% by 2020 (Google)
- 96% of Facebook users access the app via a mobile device (Google, 2020)

Mobile smartphones can access websites, as well as perform a multitude of other tasks, which is why they have become more of a necessity than a luxury these days.

For you, as a law firm owner, this provides a unique opportunity to connect with local customers via their mobile devices.

Before you start to develop a mobile arsenal to drive more inbound calls, you must first figure out whom your mobile competitors are. It is important to know whom you are up against in mobile marketing so you can plan your strategies accordingly.

To effectively do this, you need to identify your closest competitors and learn what mobile techniques they are using to generate their sales.

Mobile-Optimized Websites

First, find out which of your competitors have a mobile-optimized website. One quick and easy way to find out is to pull up their website on your mobile phone.

Did it load quickly? Was it easy to find their contact information and other details that consumers tend to look for while on-the-go? Was it optimized to fit your phone screen? If so, they have invested in their business by making sure their mobile customers and prospects are taken care of.

Now, pull up your website on your mobile phone. If it's a nightmare, it's not your phone that is the problem, it's your website. This means you have been losing potential business.

Text Message Marketing

Next, figure out which of your competitors are using text message marketing. If your competitors are doing it, they are probably telling the world to "text 123 to example." If you see promotions such as this, they are using text messaging to build a list of repeat customers.

This is one of the most cost-effective and results-oriented forms of marketing today. Text message marketing allows your competition to draw in local consumers with a great offer. Then they send out occasional messages or coupon offers to keep them coming back to use their services.

Let's say one of your customers had plans to contact your law firm today after work, but they recently joined your closest competitors mobile list and had received a text coupon offer from them before they had the chance…

Who do you think the customer will call?

There are many other forms of mobile marketing your competitors could be using to capture the attention of local consumers such as mobile SEO, QR codes and mobile apps.

If they are using these methods, it may be in your best interest to start researching how your law firm can do it even better.

Analyze Your Current Mobile Marketing Status

What is your status when it comes to staying connected with local consumers using Mobile Marketing strategies?

Researching your competition is a necessary task if your goal is to become the local authority in your specialty. But it is equally important for you to analyze where your law firm currently stands in order to move forward.

Are you currently running a mobile marketing campaign, but not seeing the results you want? Or, do you want to start a mobile marketing campaign but keep putting it off because you don't know where to begin?

Every law firm in your local area is in a crucial fight for more clienta and profits. Therefore, in order to enjoy a spike in sales, your law firm can no longer ignore the profitability of ramping up your mobile efforts.

Analyze Your Mobile Status

Many law firm owners pump a lot of muscle in competing with similar law firms, while neglecting to take a close look at what they're doing.

Analyzing your mobile status will help you figure out which weaknesses are holding you back and which strong points can help you win the war.

You need to understand where your past efforts have taken you, as well as what your future has in store for you based on where you stand today.

For starters, it is crucial that you take note of what you are and aren't doing to generate more clients using mobile marketing.

Ask yourself the following:

- Is your mobile website user-friendly? Does it load within seconds or take forever to render properly? Does your mobile website have all the relevant information on it that clients look for while on the go?
- Does your mobile website come up high in the rankings on mobile search engines, or is it nowhere to be found when local consumers perform a search for your "service + your city" on their mobile devices?
- Have you started to build a text marketing list? If so, what are you currently doing with that list? Are you focused on building a trusting relationship or are you spamming them with offers on a daily basis and getting high rates of opt-outs?
- Is your opt-in/call-to-action on all of your printed and web marketing materials?
- Are you using QR codes as an additional method of increasing awareness about your business? Do you have your QR codes on all your other marketing materials? Are you using them to direct traffic to your mobile website?
- Do you currently use a mobile app to keep your audience engaged?

As you can see, there are a lot of things to consider when

it comes to making sure your law firm is on the right track toward beating your local competition with mobile marketing.

Spy on Your Mobile Marketing Competitors

Do you want to know how your closest competitors are driving more business by using mobile marketing? Just take a look at their campaign yourself.

Mobile marketing has recently opened new doors for businesses that want to market their products and services by using mobile phones as personal "mini billboards". This has been enhanced by the fact that more and more people own mobile devices, and use them to find local products, services and businesses regularly.

To beat your competitors in the world of mobile marketing, you need to know what they are doing to be ahead of the curve. Digital technology is growing at astonishing rates and is not expected to slow down anytime soon.

This alone is causing many companies to be left behind when it comes to new-age technology.

Spying on your competitors' mobile marketing initiatives may seem like a daunting task, but it's not. In fact, all you need to do is identify which are taking most of your customers and let the research begin.

You should begin by visiting their mobile websites on your phone. Go through the websites and take note of the look and feel, the features and the traffic flow. Although your goal is NOT to copy exactly what they're doing, you could get a few pointers for your own mobile website.

Next, find out how their text message marketing campaigns operate simply by joining their mobile list. They probably have a text call-to-action placed everywhere, so opt-in and pay close attention to what happens throughout the entire process. This is the perfect way to get a first-hand look at their services, products, and promotions.

Are your competitors using QR codes to generate interest in their law firm? If so, whip out your mobile phone and scan their codes to see what lies behind them. Where do the QR codes take you? What type of incentives are they offering to get people to scan them?

Another thing you can investigate is your competitors' mobile applications. Download their apps and see what they're offering and how user-friendly they are.

The information you gain from your research should be used solely to set up your mobile marketing campaign that not only beats your competitors but also attracts new customers and keeps them loyal to your law firm.

Spying on your competitors is not illegal, but there are limits you should follow to remain fair. Under no circumstances should you use unethical measures to jeopardize your competition in your quest for mobile marketing.

Make Customers Call Your Law Firm with Mobile Marketing

The secret to beating your competitors in the legal industry is making your law firm more interesting to your target audience. There are several ways to do this using mobile marketing if you plan ahead, focus on the right things, and maintain your campaigns over time.

As much as you would like to boot your local competitors out of the picture, the fact is a lot of them will probably be using some of the same mobile marketing methods as you are.

So, your main focus should be geared toward making your clients choose your law firm over theirs. This is fairly easy to do if your efforts are consistent and persistent.

It is up to you which tools you use to work positively toward attracting new clients and keeping the ones you already have.

Here are a few tips which can work in your favor and help local consumers choose you:

- You need to have a good website that is mobile-friendly and easily accessible by mobile phone users in your area. People are using their mobile phones to access the web to search for local products and services while on the go. Make sure your site loads quickly, gives them the exact information they need, and is easy to navigate.
- If you choose to start a text message marketing campaign, make sure your text messages offer great

value, relay a clear message, and are short and informative. Also, be sure to send messages out consistently, yet conservatively. Create a careful balance that makes sense for your business and your target audience. Need a boost in getting new mobile subscribers? Give your customers and prospects a great incentive in exchange for opting-in and watch your list grow exponentially.

- Consumers love businesses who stay "on top" of the digital age. They expect you to have a website, be actively involved in their favorite social media outlets, and to be easily accessible from their mobile devices. Have a mobile app developed to aid in keeping your local consumers connected with your business. Implement the use of QR codes as a way to keep your local consumers engaged and provide them with "instant gratification."
- Mobile SEO should be used effectively to attract qualified traffic to your website. Mobile users search for local products and services constantly on their mobile devices when on the go. If your law firm does not rank in the results, there is major potential profit leak left for your competitors to scoop up.

Finding Your Business Basics

If somebody goes online, searches for your services, and gets to your website, they probably want to just get the basic information. They probably are not interested in learning a ton of information about you. They simply want to find who you are, where you're located, what your services are, and then press a button to book a consultation.

I have an example below of a mobile version of a site where they can just get the basic information, hit that "Book an Appointment" button, and then schedule an appointment. You should absolutely set up a mobile version of your site, give the basic information, and don't overcomplicate it.

Now that you have your website conversion fundamentals in order and have a proactive Mobile Marketing plan, you can start to think about Social Media Marketing.

NINE

Social Media Marketing

How to Leverage Social Media (Facebook, Twitter, LinkedIn & Other Social Platforms) for Maximum Effect in YOUR BUSINESS

There is a lot of BUZZ around Social Media (Facebook, Twitter, YouTube, LinkedIn), but how can it be leveraged by a service provider? How can you use social media to grow your law firm?

In this chapter we are going to cover social media marketing for your law firm. I hope that by now, you've learned a lot about how to position your law firm online, how to rank well on the organic listings on Google Maps, and how to rank well in the organic non-paid listings.

Now, we're going to talk about social media marketing, and how you can utilize social media tools like Facebook, Twitter, and LinkedIn to grow your business.

As I talk to law firm owners throughout the country about Internet marketing and social media, I tend to get a puzzled look. The question is, "How in the world does all of this social media stuff apply to my law firm? How can I possibly use Facebook in a way that would help me increase my revenues, boost my clients, and get more repeat business?"

I'd like to try and bridge the gap on where the "lowest-hanging fruit" for social media is in your law firm by asking, "What's your number one source of business today?"

Just stop and think, where does most of your revenue come from? You'll quickly come to the conclusion that your number one source of revenue is repeated and referral business.

= **More Repeat Business & More Referrals**

The lifeblood of any service business is your existing clients returning for services over time, and your existing clients referring you to their friends and family. If social media is harnessed correctly, it gives you the ability to take that repeat and referral business, inject it with steroids, and take it to a whole new level.

Let me explain why I feel that it's a great place for you to really connect with your clients and get more repeat and referral business. Just a couple of Facebook stats gleaned from Google:

- Facebook currently has 2.37 billion users
- The average user has 338 friends
- Users check in an average of 14 times per day

If you can get your real customers, current and past (your sphere of influence) to connect with you on social media, such as Facebook and/or Twitter, your law firm is exposed to their 388 friends as soon as they "like" and follow your page.

It's almost as if they'd sent an email, or they'd sent a text message out to all their friends saying, "I recently received a service from this law firm in our area. The next time you need their services, why don't you think about them?" It's extremely powerful to gain exposure to their sphere of influence.

Another major advantage is that they've given you permission to remain top-of-mind with them. The average user, like I said, checks in 14 times per day. They login to check out the updates on their Facebook wall and to see the updates of all the companies and people they have liked or are friends with. If you're posting updates to your social media profiles, the people who have liked your page are going to see that new content whenever they login.

They are going to see an update and your logo. They're going to see some special offer or promotion, and it's going to pique their interest. Next time they need your services, who do you think they're going to call?

Top of Mind Awareness

There is a higher probability folks who have liked your page are going to use you again, and refer you to their friends, because they remember you and had a good experience with your service offering. They know who you are.

You've remained top-of-mind. If you look at major companies like Coca Cola, Pepsi, and Lay's, they spend billions of dollars a year on advertising and promotions, through TV, radio, and print.

What's the whole thought process behind that? They're

developing their brand, so they can maintain what we call "TOMA," top-of-mind awareness. Leveraging social media inside your existing sphere of influence is a great way to tap into that top-of-mind awareness.

Where should you start? Where can you start using social media, with all of the different platforms out there? With so many different social media tools, what should you be using?

In chapter two, we talked about having a blog and putting out consistent updates. Well, blogging ties very nicely to your social media strategy. These are the social media profiles you definitely want to have set up and ready to roll in your business.

Finding Your Followers

Let's talk strategy before we get into the granular details. Talk about high level. How do you leverage social media and how do you gain that initial following?

Well, first, you want to utilize email to get initial engagement. Having an active social media profile with daily updates is not worth a hill of beans if you don't have likes or views.

Now, at the same time, if you have thousands of irrelevant people that have pressed like on your website or on your Facebook profile, it's not going to work to your advantage if they're not people in your service area. They're not the target market that we discussed in the marketing fundamentals.

You want to make sure that you have a strategy to get your real clients and your true service area engaged with you in social media. You should leverage email to engage your clients to get to your social media profiles. This takes a multiple step process.

The first thing you want to do is build that list or go into your customer relationship management system, if you have one, and export the name and email addresses of your clients. Include current clients, past clients, sphere of influence of your friends, your business partners, the people that you do business with, and put them into an email.

Queue up a nice little message, which says, "Hey, we appreciate your business. We appreciate your relationship over the years. We're getting active in social media and would love to have you engage with us. Please go to Facebook.com and press the Like button." Make sure to give them a direct link to your Facebook page.

There are a couple of things you can do. You can offer them an incentive, something of value like a coupon or a discount. Or, if you feel like you've got an active customer base that knows who you are and likes you, just ask them to do it as a favor.

You'll be able to start building that following. Now, you don't want to stop there. You don't want to just send one email out that says, "We're on social media." You now want to build it as part of your law firm.

Just Ask!

In the Google Maps Optimization chapter, I talked about having an email go out after service, thanking the customer for their business and asking them to go ahead and write a review for you on one of the various online directory sites.

Well, there's no reason you couldn't send a subsequent email to that contact, maybe a day or two later, which says, "By the way, we're actively involved in social media and would love it if you would engage with us." Then give them a direct link to your social media profiles where they can press like, subscribe, and follow to start engaging with you on social media.

The key is that it needs to be an automated process where you're typing your customer's name and their email address. These emails go out to everybody that you serve without any hiccups, and without any potential for dropping the ball. If you don't do it consistently, you won't get a true following - and you won't get your real customers engaging with you on these social media platforms.

That's step one. Leverage email to build that initial engagement and following of your real customers. Remember, we want authentic customers, not just throwaway links and subscribers.

Once you've got that part squared away, you have got to think about what you are going to post. What information are you going to put up and how frequently? You should post to your social media profiles once a day. If that seems like too much for your business, post once a week at a very minimum.

What to Post (and Why)

These should be informative posts. It should not be a sales pitch.

You can do that every now and then but more than 80% of the time it should just be social content: "What to know if you've been injured in a car accident", "How has COVID impacted the courts", "Here's a picture of us at our team building.", etc.

Keep it informational, keep it relevant, keep it social, and then you must engage. Social media isn't a one-way dialogue. You shouldn't be going to your social media profiles and pushing out updates that don't have any engagement. You shouldn't just be posting. You should be trying to get people to reply to your post: "Hey, that was funny", or "That's a beautiful picture", or "Thanks for that great tip," all of which you can reply back to.

Then, listen to what your fans are saying. Once you've got a flow – you've got 50, 70, 100 or a couple of thousand people that have liked you – you are going to be able to hear what they are saying as well. They might post something that's totally irrelevant to you, like "Hey, tomorrow's Billy's birthday." There is no reason that your law firm couldn't reach out and say, "Hey, wish Billy a happy birthday for us!", from your law firm. They will think, "Wow, this is a firm that cares. This is a firm that's real and authentic."

Engaging in social media is probably a lost art. Most people that use social media just post one-way messages, which is not the idea. It's a social platform, so there should be conversation. There should be dialogue.

Fill in the Business Bio

The next thing you want to do is to develop your brand and make sure that you enhance the bio section on each one of these profiles. Within Facebook, Twitter and LinkedIn, you will have the option to fill in an 'About Us' or bio section. Write some interesting information about your firm there.

Take the information from the 'About Us' page on your website where you talk about where you were founded, why you started the firm, the service that you offer, etc., and pop that into the bio section on your social media profiles.

You also have the ability to put an icon on each one of these social profiles, and you want to make sure that you're using an image that represents your business. It can either be a head shot of the owner or it can be a logo.

Below is an example of a few social media icon options that go from good to bad to a big no-no!

Good
Company Logo, even better if you are a well known brand

Good
Mary Thompson
President
Mr. Rooter

Bad
Glad you like shaggy, and maybe it's your knick name. But it's not relevant to your business

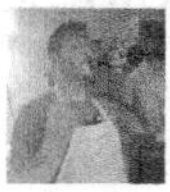

Very Bad
Drinking a beer while giving the middle finger, unbelievably common, but a poor choice

If your personality represents your brand, then it might be a good idea to use a nice head shot so that people connect and resonate with you. People tend to buy from individuals more than they buy from businesses because a business is an anonymous entity, and a person is someone that they feel they can get to know, like and trust.

Don't be like our bad examples, "Shaggy" or the "drunk contractor." Don't put a picture of yourself in a T-shirt with a beard grown out. Be professional. Represent yourself as an important part of the business. Stick with the examples on the left – the logo and/or the professional head shot, as opposed to Shaggy or a weekend photo of you doing something crude and lascivious.

It's all about branding, so make sure that you're leveraging the header graphic and the image icon. If there is an option for you to customize the background, do it! You want to make sure that your elements marry up with the overall branding of your firm.

Make sure everything on your social media profiles is consistent with your website. On your website, you've got a color scheme, a logo, and maybe you have printed brochures. Be sure to keep a consistent flow, look, feel, and color scheme on all your social media profiles, website, and offline materials.

Posting Plans and Pointers

Don't forget to have a plan for social media. How often are you going to post? What types of posts are you going to put out there? Who is going to be responsible for posting them? How are you going to engage your customers? Which social media profiles is your business going to be involved with?

Remember in chapter two we talked about the fundamentals of your marketing plan (market, message and media). Be sure that you have a clear understanding of who your client is and who your ideal client is. Then make sure that you are crafting a message that will resonate with that particular client. It's important to consider all of these things as part of your social media strategy.

Don't just dive in. A common mistake would be to just setup the profile and start posting with no thought process or plan behind it. Think about it. What pages are you going to be on? What message are you going to put out? What color scheme are you going to use? Set all of that up and then get very specific about your target. Is your client the commercial type? Is your client a residential type?

One solid method is to schedule your post types on specific days, such as:

Monday, Wednesday and Friday are the days that you are

going to put up tips. For example, DIY tips are a great place to get started and you can dish out expert tips related to your service offerings, etc.

Tuesday and Thursday, you'll post photos, such as pictures of really interesting things relative to your business, pictures that showcase your work effectively, before-and-after pictures of a job well done. Pictures that are interactive are also good choices.

Saturday and Sunday, you post coupons.

I am not saying this is the editorial calendar you must follow. This is only an example. However, the point is to make it easy for yourself so that you know what is going up and when. Your posting process can be streamlined, and it can also be automated.

Leveraging Posts

When we talked about the blog in the SEO chapter, we went over leveraging content. Because content is king, you have to be creating updated information on a consistent basis.

This content can go up in various places. As you post a new piece of content, it can go to your Facebook and Twitter pages automatically. It can go straight out to Pinterest if it has a photo included, and you can take your blog content and syndicate it to recreate great social media content.

Remember, content isn't necessarily just written text. You are an expert in your craft. You know things that the average consumer doesn't, such as what to do in the event of an emergency, why somebody would want to consider a specific service offering versus another, or why somebody would want to consider a new service offering you have recently started.

You can either sit down and write about it, you can take an audio recorder and record yourself talking about it, or if you're comfortable on video, you can break out the camera phone and shoot a video talking about an issue your ideal client may be facing.

How 1 Equals 5

Once you make a video, you can get a lot more bang for your buck. That one piece of content can serve multiple functions. The first function can be posting videos up on social media or on websites where you can upload interesting clips and videos like YouTube or Vimeo.

One video post can result in several forms of content.

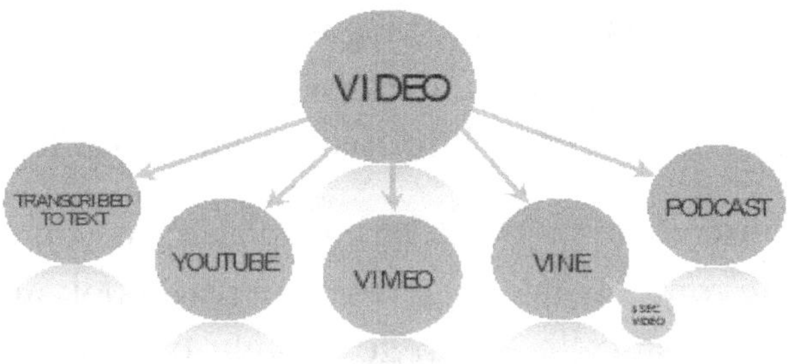

You can also have the video transcribed using a service like castingwords.com. There are various transcription services available.

That video of you talking about the benefits of your service offering can now be transcribed into text, which may then be used as a blog post and be syndicated into your social media profiles. Another step beyond that is using that same audio and turning it into an audio podcast you can have hosted on your website.

There are a lot of things you could do to take your content and work with the modality that you're most comfortable with. Some people like to write. Some people like to talk. Some people like to be on video. Figure out what you are most comfortable with and run with that. This is how you create social media content for your online marketing plan.

Remember, educational content that's published in multiple places gives you industry expert status. By publishing and getting picked up in industry listings, the local newspaper or a reputable blog, you are considered an expert. This is going to drive your credibility, which in turn, will result in more referrals.

I want to give you some examples of good social media posts.

The following example is a great tip for English spelling. It's, "Here's a look at 11 weird, random facts about English spelling. We're not sure this will make our jobs as expert authors any easier, but it's a nice distraction. https://bit.ly/2OyfV7c."

Then, of course, a shortened link back to the website. The point is to give a quick tip like that once a day, keeping you top-of-mind, and helping you to put out fresh, relevant content.

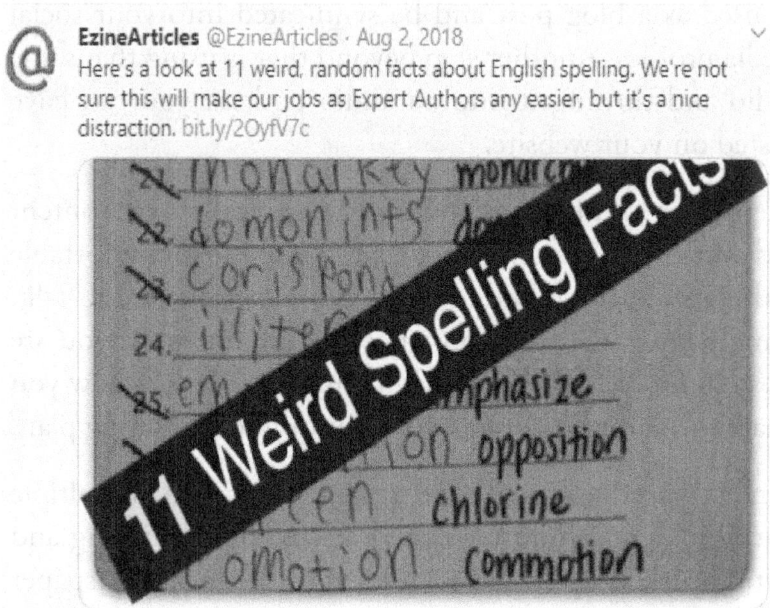

Here's another example. You wrote a post, and you're using a tool like GrapeVine. It can grab the RSS feed and automatically have that post hit your social media profile. If there's a new blog post, you can post something like "The ultimate guide to virtual reality," and of course a link to where the blog is posted.

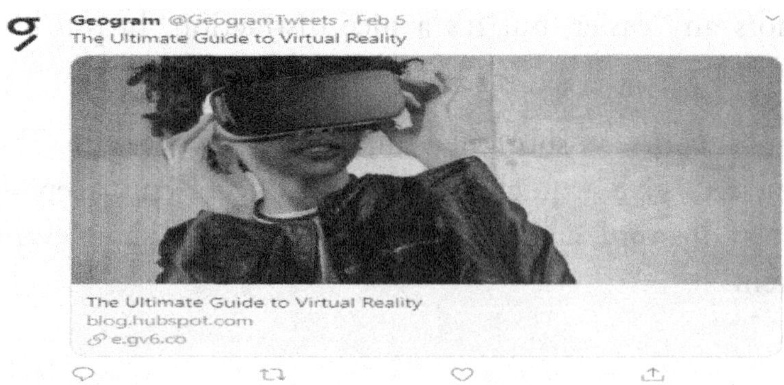

Here's an example of how you can do something interesting with pictures. In the picture below an interesting quote is being used. It says, "Life is short. Smile while you still have teeth." You'd be amazed at the things you can do with pictures. There's 200 percent more engagement with photos on Facebook than there is with text.

It's easy to find an interesting picture like this, post it with a quick little quip, and get some great engagement from your followers on social media.

Another sample, to wish your users a very Happy Weekend. Simple, but effective.

Do's and Don'ts of Posting:

1. Use the 80/20 rule for marketing messages. Put out 80% information and 20% marketing.
2. Keep it business related. Your political and religious beliefs are never a good mix with business.
3. Photos of your kids playing tee ball are good, but don't let it dominate your page.
4. Keep your vacation photos on your personal social sites.
5. Keep your business opinions, beliefs, and interests to yourself.

Sometimes knowing what not to post is more important than knowing what to post, because the natural tendency is to go to these social media profiles, and just post promotional material.

So, don't post a coupon every single time you log in. If you do that, everybody who liked or subscribed to your page will start to disappear before you know it. They'll stop subscribing, they'll unlike you, and they'll unfriend you.

You have to use the 80-20 rule for messaging: 80 percent informational and fun stuff, and only 20 percent should be promotional.

Keep it business related. You don't want to get into your political and religious beliefs, because if somebody disagrees with you, you can create a negative atmosphere.

That's not something you want to do on your business profiles. You've got a personal profile for a reason. If you want to put your religious and political beliefs there, knock yourself out. Just keep it off your business pages.

You may not necessarily want to put too many photos of your kids on your page, even if you're a personality brand. Don't let your kids and your family be the dominant positioning behind your business profile page.

Obviously, keep your vacation photos and again, your opinions and beliefs, off your business page. Family photos are another thing that should be kept primarily on your personal profile.

When and How to Engage

We talked about asking your clients to 'like you' on Facebook and asking your customers to write testimonials. We also talked about interaction and responding to your clients actions. "Hey, thanks so much for the follow. We appreciate it." Or, if they write you a testimonial, make sure you blow that up.

Not only should you say thank you, but you should also share it. "Hey, Jean, thanks so much for the positive testimonial. We appreciate your feedback. We appreciate your business, and this is what keeps us going. This is what we're in this business for."

Then, you could take the testimonial and put it on your website or embed it on your website through the various widgets and short codes that Facebook provides.

For example, you create a post saying, "Seniors should live at a place where they can maintain their normal lifestyles and follow their interests. Let us know how we can help." Then, one of your followers says, "Wonderful caring staff. So lovingly devoted to those they serve." Then you respond: "Thank you for your comment! Our office phone number is 254-1799. We look forward to answering any questions you may have!" This is your engagement. You want people to make comments, and then you want to be able to talk back.

This is just a way to put out relevant content, and if you're paying attention to your feed, you can turn it into some great conversation. Again, you want to be active on social media.

It's a great way to get repeat or referral business. You need to be on Facebook, LinkedIn, Twitter and YouTube. You want to utilize email marketing to gain that initial following, and then post updates that are informative and not sales-oriented on a consistent basis and engage.

If you do this regularly and correctly, you're going to grow a nice following of real customers in your true service area. You're going to remain top-of-mind and it's going to help you grow your business in terms of the lifeblood of your organization, which is repeat and referral business.

Top Twitter Abbreviations You Need to Know

Technical Twitter abbreviations:

• CC = Carbon-copy. Works the same way as email
• CX = Correction
• CT = Cuttweet. Another way of saying partial retweet
• DM = Direct message. A direct message is a message only you and the person who sent it can read
• HT = Hat tip. This is a way of attributing a link to another Twitter user
• MT = Modified tweet. This means the tweet you're looking at is a paraphrase of a tweet originally written by someone else
• PRT = Partial retweet. The tweet you're looking at is the truncated version of someone else's tweet.
• PRT = Please retweet, a plea to put at the end of a tweet
• RT = Retweet. The tweet you're looking at was forwarded to you by another user

Industry Twitter Abbreviations:

- EM = Email Marketing
- EZine = Electronic Magazine
- FB = Facebook
- LI = LinkedIn
- SEO = Search Engine Optimization
- SM = Social Media
- SMM = Social Media Marketing
- SMO = Social Media Optimization
- SN = Social Network
- SROI = Social Return on Investment
- UGC = User Generated Content
- YT = YouTube

Common Hashtags and Chats:

- #BrandChat = private chat about branding
- #CMAD = Community Manager Appreciation Day
- #CMGR = Community Manger topic chat
- #FB = The user is sending this post to Facebook
- #FF = Short way of saying Follow Friday, or a recommendation that others follow the user
- #in = the user is sending this post to LinkedIn
- #LI = This user is sending this post to LinkedIn
- #LinkedInChat = For general use questions and questions about marketing/self-promotion on LinkedIn
- #Mmchat = Marketing and social media chat
- #Pinchat = a chat for maximizing Pinterest use
- #SMManners = Social media manners chat
- #SMMeasure = For discussion of analytics and measurement

- #SMOchat = Social Media Optimization chat
- #SocialChat = Social media chat
- #SocialMedia = an all-inclusive chat for subjects big and small in the subject of Social Media

Example of a Customized Facebook Page:

Example of Branded Twitter Profile:

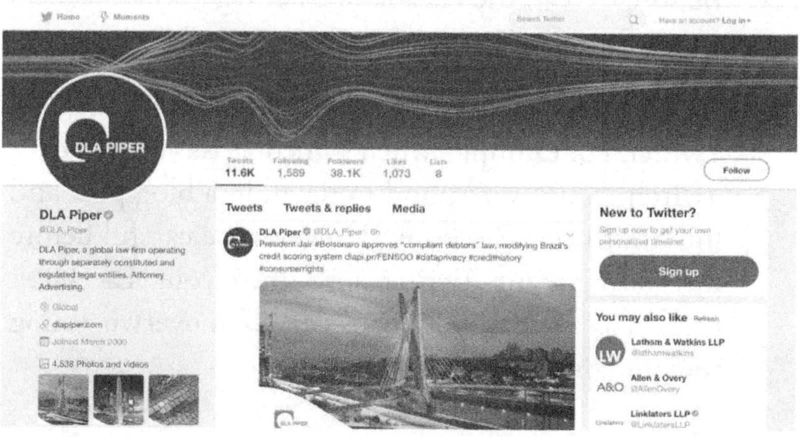

How to Use Facebook for Business

There are so many ways and techniques, which if applied with proper sense can change the fate of your business. To get started:

- Fill out your profile completely to earn trust.
- Establish a business account if you don't already have one. It is similar to creating a personal account. The page looks like the following:

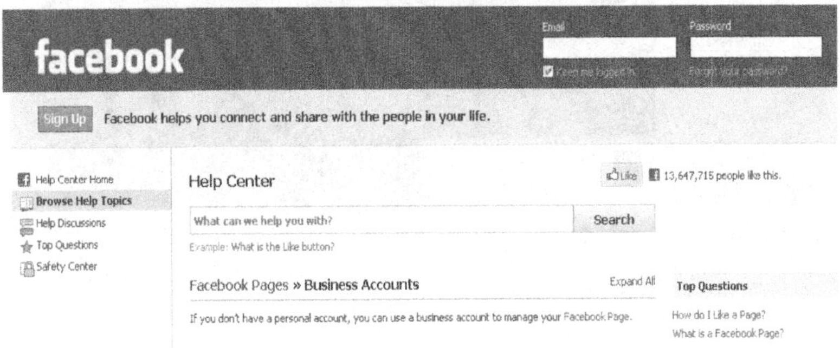

- Be sure to read the Facebook rules regarding business accounts.
- Combine Facebook with other social media tools like Twitter. For example, when someone asks question on Twitter, you can respond in detail in a blog post and link to it from Facebook. However, be careful before integrating your Twitter feed into your Facebook profile, as a stream of tweets can seem overwhelming to your contacts.

- Keep any personal parts of your profile private through Settings.
- Create friends lists such as "Work," "Family" and "Limited Profile" for finer-grained control over your profile privacy.
- Post a professional or business casual photo of yourself to reinforce your brand.
- Limit business contacts' access to personal photos.
- Post your newsletter subscription information and archives somewhere in your profile.
- Obtain a Facebook vanity URL so that people can find you easily.
- Add your Facebook URL to your email signature and any marketing collateral (business cards, etc.) so prospects can learn more about you.
- Post business updates on your wall. Focus on business activities, such as "Working with system technologies on web site redesign."
- Share useful articles and links and valuable resources on your wall that interest customers and prospects to establish credibility.
- Use Facebook Connect to add social networking features to your web site.
- Suggest Friends to clients and colleagues. By helping them, you establish trust.

- Buy Facebook ads to target your exact audience.
- Start a group or fan page for products, services, your brand or business. Unless you and your law firm are already household names, a group is usually the better choice.
- Join network, industry and alumni groups related to your business.

One of the biggest benefits that social networking platforms like Facebook provide the business community is the ability to get repeat exposure with the people in your network.

Clients, peers, and prospects can make up your network, and you can promote events, sales, special offers and more through your Facebook profile.

These are very common and essential tactics to implement to use Facebook for your law firm.

How to Avoid Mistakes on Facebook

With more than 500 million users, Facebook has become a must-use marketing platform for businesses of all sizes. While Facebook's staggering membership stats alone are enough to entice small business owners, few actually know how to do so effectively.

Here are 10 of the most common Facebook marketing mistakes law firm owners make and how you can avoid them.

1. Not having a clear marketing purpose: Whether you've created a page for Facebook or are still working on it, now is a good time to evaluate what you want to get out of it. So set clear goals at the very beginning.

For example, suppose you are hoping to attract 500 new fans who could become potential customers in a six-month time period. Be sure to assign someone within your company to maintain the page. It's important to regularly post fresh content in the page. If you've already created your page, but it hasn't been updated in a while, be sure to update.

2. Not knowing the difference between a personal profile and a business page: There are so many major differences between personal and business pages on Facebook. You should know what they are so that you stay safely within Facebook's Terms of Service.

A personal profile is the type of account an individual shares with friends and family. But a Facebook business page is used by brands and companies for promotion purposes.

Another important distinction: If you have a Facebook page for your law firm, you have fans. If you have a personal profile, you've got friends. So, don't ask your potential customers to become your "friend" on Facebook. They need to become a fan.

3. Not understanding how your customers use Facebook: Many law firm owners don't know how their clients interact with Facebook. When you log into your Facebook account, for instance, the first thing you see is your news feed. You don't see the pages that you've liked.

It's important to understand that marketing your law firm page is not the same as posting a status update to your friends through your personal Facebook account. If you want to get your fans to view your content, you need to post directly to the Facebook page for your law firm. Your posts will show up in your news feed and will also show up in theirs.

4. Not getting the right URL for your Facebook page: You also need to know how to obtain the proper URL for your law firm page on Facebook. There are so many law firm owners who don't understand that if you have 25 fans of your law firm page, you are eligible to obtain a URL for your page that has your law firm's name in it.

For example, it could be www.facebook.com/yourcompanyname instead of the former random number URL assigned. You can go directly to www.facebook.com/username and log in to choose your URL.

5. Not responding to wall posts: It is a smart idea to remember that Facebook is all about interaction. It's very important you quickly respond to potential customers if they post questions on your wall.

The faster you can answer their question, the higher the likelihood of converting that potential customer into a paying one. You can adjust your Facebook page settings to notify you via e-mail whenever a new post is made on your wall.

6. Creating fake Facebook user accounts to boost fan count: It is a long-term process to get potential clients and existing ones to become fans of your law firm Facebook page. Although you may want to see your fan count increase rapidly, don't give in to the temptation to create fake Facebook accounts and then become a fan of your law firm page using that particular account. Facebook can often detect fake accounts, which are a violation of their Terms of Service. If you're caught, you will lose your page and the marketing power that comes with it.

7. Not updating the page regularly: You should not create a page for your law firm and let it become inactive. You should update it with fresh content at least two times a week. This will help keep your fans engaged and interested in your brand.

8. Posting poor quality content: Nothing irritates a potential client like poor content. Consumers are always looking for useful information, not hype. Before you post anything, you should ask yourself how your content benefits your fans. If you can't answer that question, revise before you post it.

9. Not using Facebook's free tools: One of the great things about Facebook is that it offers several free tools to help law firm's gauge how well they are using their page to attract and engage customers. There are many free tools in Facebook that provide law firm owners with detailed metrics about the effectiveness of their page content, analysis of user growth and demographics, and other concerns. You can find out more about it by logging on to https://developers.facebook.com/docs/platforminsights/page.

10. Not properly promoting your page: Many law firm owners aren't aware of the ways in which they can promote their Facebook page and attract potential clients. Facebook offers an advertising option that allows you to purchase relatively inexpensive ads.

You can also purchase a paid sponsorship of your post. For

example, if you've written a post about the best strategies for designing a good website, you can promote that post on Facebook through its advertising and sponsorship platform. You can also create a link to your Facebook page on your law firm website.

How to Set Up a Facebook Business Profile

As one of the top social networking websites, Facebook provides you an opportunity to create a business profile. It is another method for your law firm to reach its marketing goals.

With a business page, you can create advertisements that appear on targeted Facebook pages and allow you to reach a wider audience. So now with just an email address, you can quickly set up and customize your own business account on Facebook.

Here are a few steps detailing how to create your business profile in Facebook:

1. Access the Facebook website and click the "Create a Page for a celebrity, band or business" link at the bottom of the "Sign Up" section.
2. Choose the "Business or Brand" option.
3. Type in the page name you want to use and select your business category.
4. Add pictures.

5. Create a username for your page. (e.g. the name your customers will use to search for you)
6. Add your law firm details, such as the "about" section.
7. Add additional keywords that people might use to search for your service offerings.
8. Fill in all details; contact information, location, hours of operation and other information you want visitors to know.
9. Tell your story – e.g. what sets you apart from your competitors, how you got started, how long you've been in business, etc.
10. Create content for your page, before asking your customers to "like" your page.

Top 15 Acronyms Used in Facebook

Acronyms on Facebook are borrowed directly from decades of net culture, developed as easy Internet shorthand on electronic bulletin boards, IRC, IM, and email. Facebook's integration with SMS naturally means that texting lingo is inherently part of the argot.

As we will need to share our messages within a small word count it can be difficult to express our views with just a few words. Sometimes it may be not possible to express within that limit if we use normal language.

Although we can write complete words or sentences, using acronyms can save space. Here are 15 common shortcuts to give complete descriptions. These shortcuts can simplify sharing your views and ideas.

- 2moro – Tomorrow
- 2nite – Tonight
- BTW - By The Way
- B4N - Bye For Now
- BFF - Best Friends Forever
- FWIW - For What It's Worth
- GR8 – Great
- ILY - I Love You
- LOL - Laughing Out Loud -or- Lots Of Love
- OMG - Oh My God
- THX – Thanks
- TYVM - Thank You Very Much
- RTM - Read The Manual
- NP - No Problem
- ISO - In Search Of

Because of short message chatting and wall posting, Facebook encourages extensive use of shorthand typing. Most of these acronyms might be familiar to regular Internet users but some are very specific to Facebook.

How to Create Your Following in Facebook

Once the business page is complete, most people are anxious to get lots of followers. The truth is it takes time to build a new fan base from scratch.

Efforts to build your base should begin the day you set up a Facebook page. To create a strong fan base requires an ongoing commitment to brand, monitor, and network with people who find interest in your product. Besides quality service, it's important to build close-knit relationships with visitors.

There are some steps that if successfully applied can increase your following:

1. Be Prepared with Quality Wall Posts and Consistent Engagement

If you want to be liked, be likeable first. A disorganized Facebook page can turn off customers instantly. When reviewing a Facebook page, quality content and active engagements are great first impressions.

Several other factors people look for before joining a page include the brand itself, consistent posting of fresh information, and active engagement from both fans and admin.

2. Reward Your Loyal Supporters

You may have just started your Facebook page, but your law firm is well-established. **Encourage your loyal customers to join your Facebook page as supporters, reward them** with customizable badges/tabs (to be placed on their profiles for visibility) and special deals for consistent support. A shout-out from a happy client is a lot more attractive than a marketing slogan, creating irresistible appeal for that 'Like' button.

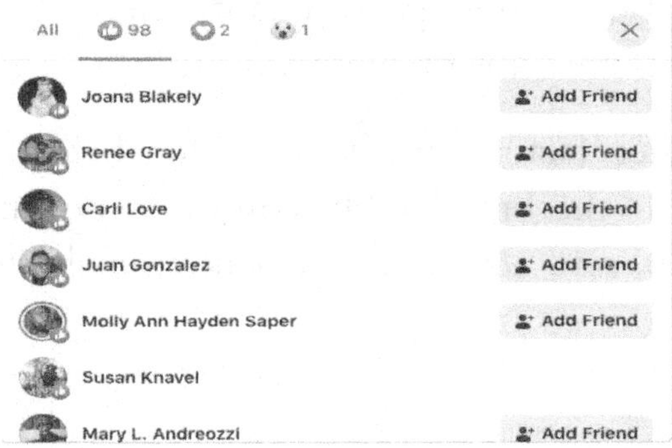

3. Leverage Your Existing Social Networks

If you've built a strong Twitter network for your law firm, utilize it to promote your Facebook page. Some people prefer not to overlap similar social contacts on both accounts, but why diminish your chance to be noticed? **Your followers can broadcast your message on both of their social platforms** by reaching out to a greater audience about your law firm.

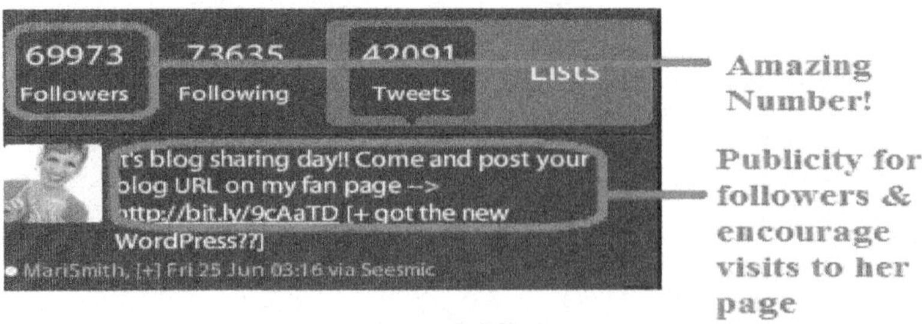

4. Integrate Facebook Social Plugins to Your Website

It's essential to have a main hub correlating all your social media activities. Your law firm's website is the only place that gives you full control over content and brand management. Integrate Facebook social plugins to encourage connections such as Facebook's "Like Box", "Like" button, and Comment stream.

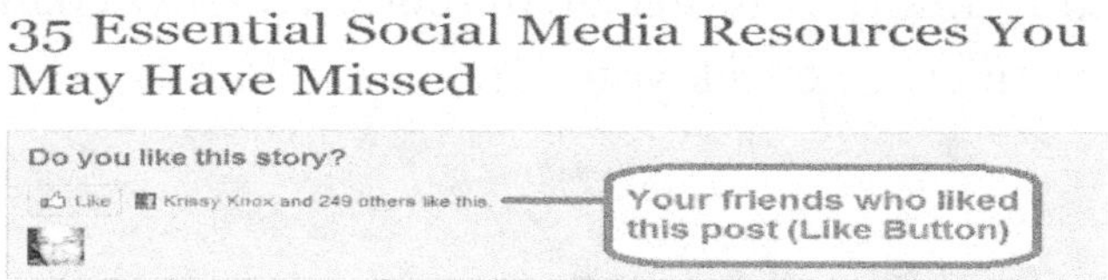

5. Remind Your Fans to Like and Share

Facebook has some easy ready-made sharing buttons with which people can promote your tabs and pages to their friends. Place a shout-out or reminder to 'Like' your status updates and **instruct fans to click that little 'Share' button right next to your message** so their friends will be alerted about the update.

6. Utilize Forum Signatures and Membership Sites

If you're an active participant in a forum or membership site, placing a signature with your fan page link is a plus. No one will care about your information unless you stand out from the crowd.

Be an active helper in a LinkedIn Group or a frequent poster of special tips and tricks. As long as your participation in the niche community is appreciated, there's a higher chance for other members to check you out.

7. Take the Initiative: Request Help From Friends

It's difficult to start a fan page with no engagement whatsoever. Why not initiate messages to your friends and buddies who are supportive of your business?

Ask them to help in some discussions, reward them with publicity or return the favor. It's easier to ask a friend than a stranger if you're worried about spamming people.

Make sure the question is interesting enough to get them talking. If you use your personal account and fan page strategically, you'll discover a huge advantage of getting new friends to be your fans while they're getting to know you better.

8. Use Tagging and Acknowledgments

A great networking tool, status-tagging, can even drive in new connections. **Tag an author or a popular Facebook page to draw attention,** but only if you have good reason to do so.

For example, selflessly promote a niche-post and how it benefits people who like your page. Be authentic, and the page admin (hopefully the fans as well) will appreciate you for it.

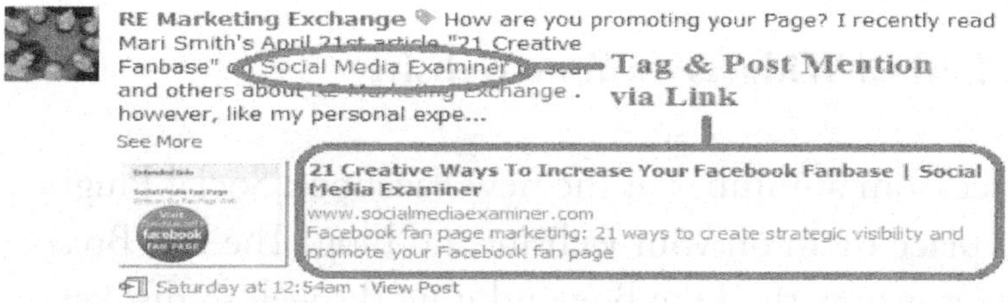

9. Participate Outside Your Page

Use the Facebook Directory and Facebook Search to **locate other Facebook pages in your niche and look for public discussions based on search terms related to your law firm**.

Provide value to the popular pages; build credibility and relationships with the admin and members. Get to know them better before asking them to look at your page. They just may reward you publicly.

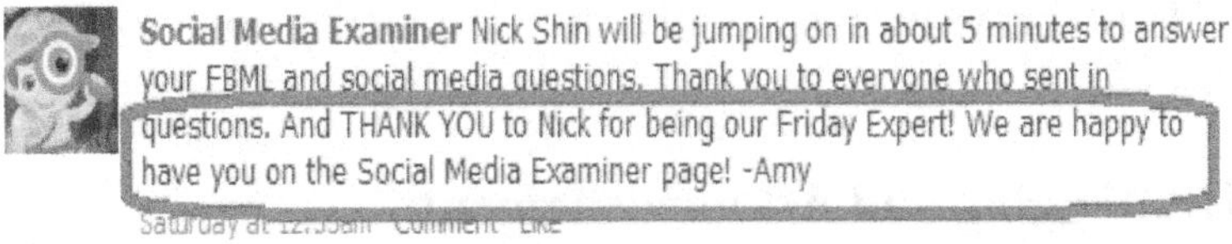

10. Collaborate with Other Page Admins for a Social Event

You can collaborate with other page admins to create a special event that may benefit both your fans and bring in new connections. I find this to be very successful. There should be a mutual understanding and proper planning to make it work for everyone's professional goals.

11. Embed Widgets on Your Website

Select from a number of the new Facebook Social Plugins and place them on your website and blog. The Fan Box widget is now the Like Box and it works well to display your current fan page stream and a selection of fans.

12. Make a Compelling Welcome Video

Create an attractive landing tab (canvas page) with a video that explains exactly a) what your fan page is about, b) who it's for and c) why they should become members. The result: you'll increase your conversion rate from visitors to fans.

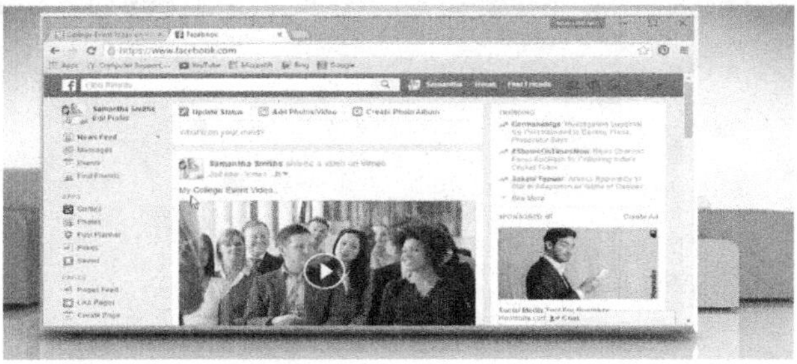

13. Use Facebook Apps

The app adds a tab to your fan page called "Shows" and when you broadcast as your fan page, everyone can view by default. (You can also broadcast as your personal profile and selectively invite friends/friend lists.)

By announcing via Twitter, your personal Facebook profile, your blog and your email list, you can broadcast regular live Internet TV shows from your fan page and create buzz.

14. Integrate the Facebook Comment Feature

You can also post any of your products as well as **Like and Comment on any item and choose to have that comment posted to your Facebook profile**, as shown in this screenshot:

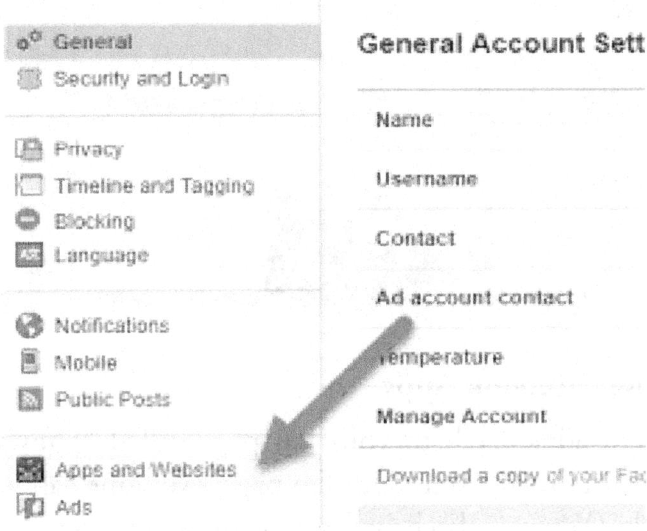

15. Place Facebook Ads

Even with a nominal weekly/monthly budget, you should be able to boost your fan count using Facebook's own social ad feature. It's the most targeted traffic your money can buy.

To buy an ad, scroll to the foot of any page inside Facebook and click the link at the very bottom that says "Advertising." From there, you can walk through the wizard and get an excellent sense of how many Facebook users are in your exact target market. When you advertise your fan page, Facebook users can become a fan (click the Like button) right from the ad as shown in the screenshot below.

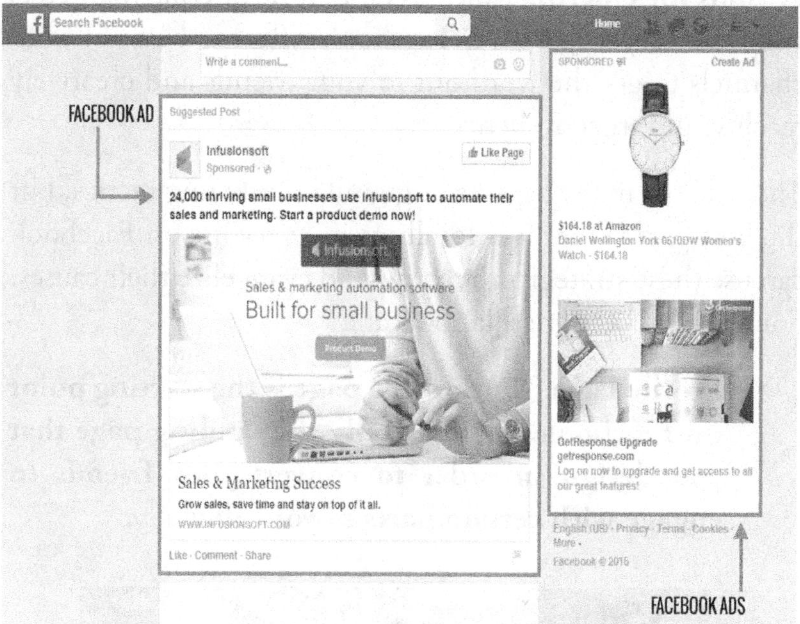

How to Build Your Brand and Market on Facebook

A large portion of web users today are spending their time and attention on Facebook, and this includes law firm owners. However, most of the marketers lack a comprehensive knowledge about how to implement Facebook for marketing their brand.

It is true that like any other social media platform, Facebook also demands some strategies to be performed to brand your business page.

Facebook offers many tools and apps to build a strong brand. Here are three chief tools to know and understand:

1. Tools for Guerilla Marketers: For the guerilla marketers who are very aggressive, Facebook offers a bevy of viral channels to get the word out to your friends and creatively reach your target audience.

There are so many aspects of guerilla marketing tactics, but the best part is that it is totally free. Everyone on Facebook can use these strategies to recruit and evangelize their causes. Some of the best guerilla tools are:

- **Profile Page: Your profile page is the starting point on Facebook. It is basically the landing page that you design in order to convert your friends to engage with certain parts of your identity.**

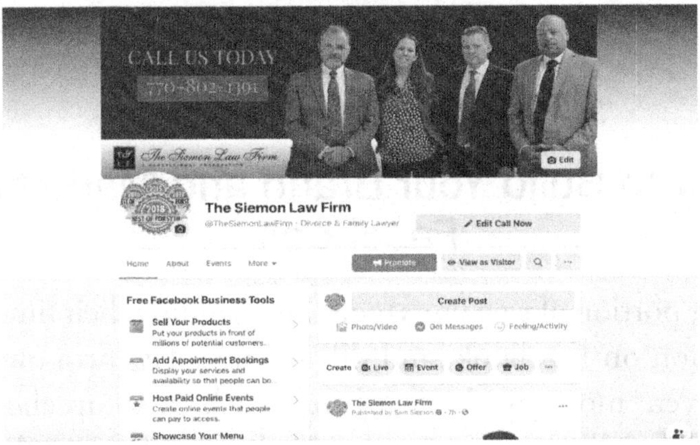

Your profile page is an opportunity to craft a credible real-world story around the reasons your products or services are so valuable. Take advantage of Personal Info, Work Info, Photos, and applications to tell bits and pieces of your narrative as it relates to your brand.

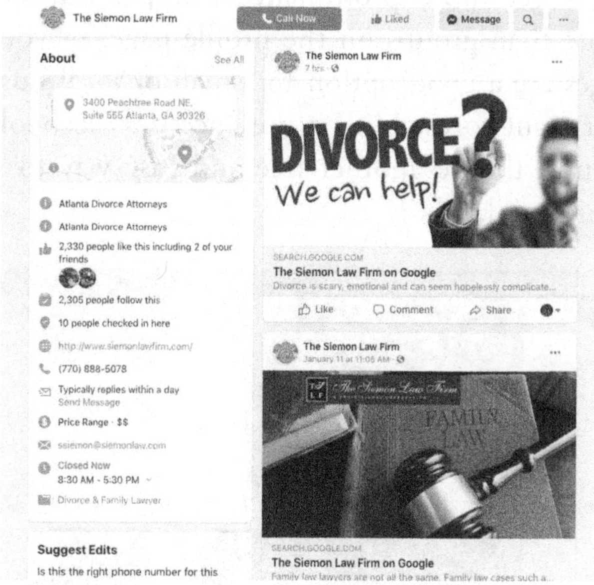

- **Facebook Groups:** Groups are the oldest and simplest way to build community around your brand or company on Facebook. By starting a group, you create a central place for customers, partners, and friends to participate in conversations around your brand.

  Here you can post discussion topics, photos, videos, and links right out of the box.

- **Facebook Pages:** Pages were launched by Facebook as a way for businesses of many types to easily establish a brand presence on Facebook.

  Pages are more customizable than groups. You can add HTML, Flash, or even Facebook applications to your pages to extend their functionality and the depth of experience users can have with your brand.

Pages get more prominent "Bumper Stickers" real estate than groups on the profile pages of your fans. Pages are a good option for small or local businesses that want to establish a presence on Facebook. Like groups, they're another free and easy way to market your brand.

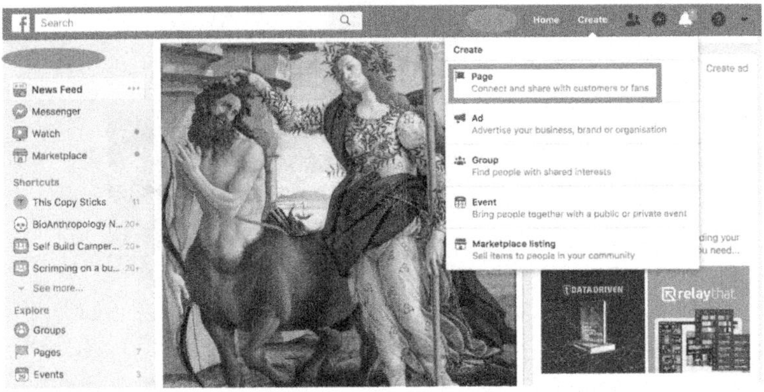

- **Facebook Events:** Facebook Events is a free application developed by Facebook that anyone can use to promote marketing events, sponsored parties, or even product launches, transactions, or company milestones.

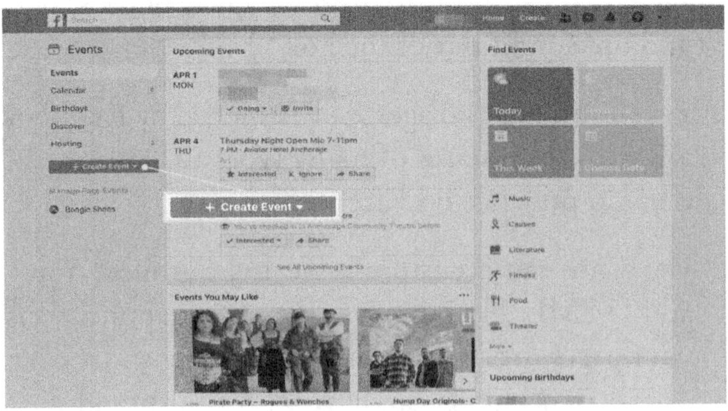

- **Facebook Notes and Photos:** Notes and Photos are two Facebook applications that allow you to share blog posts and pictures with your friends. You can use these features to post content about your brand but be careful to always do it authentically – don't be spammy.
- **Facebook Messages: Facebook** messages can be a powerful vehicle for targeted marketing on Facebook. Messages are like email, except a lot less fully featured – Facebook offers no way to search, sort, filter, categorize, or star messages.

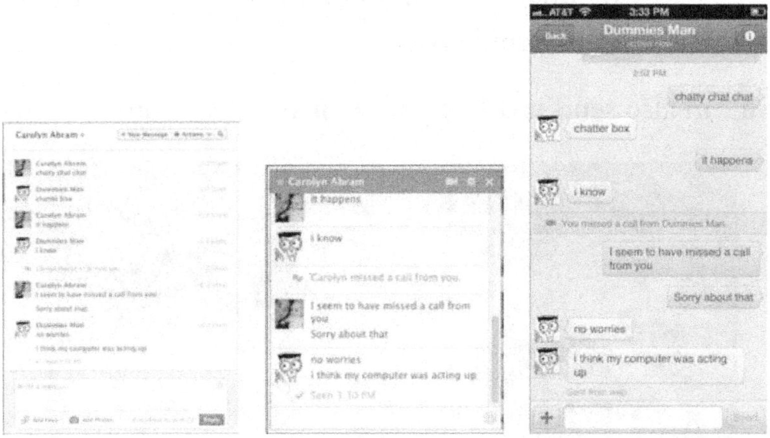

- **Facebook Marketplace:** Marketplace is Facebook's classifieds listing service. You can post a for sale ad or wanted ad in any of your networks for free. Unlike other Facebook-developed applications, Marketplace does not get heavily used by most members.

  However, unlike Craigslist, which is anonymous,

all Marketplace responses are tied to real Facebook accounts. When you receive a response to your Marketplace listing, you can see the respondent's profile page even if they're not your friend.

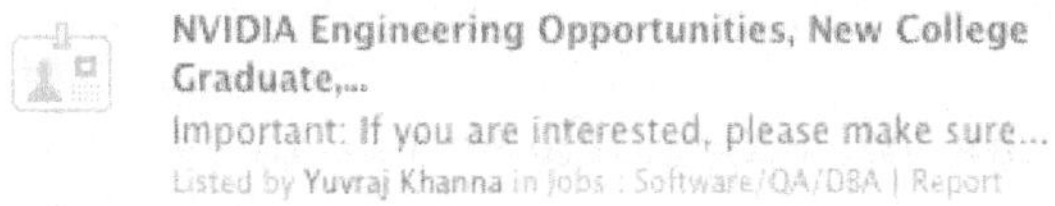

- **Facebook Share / Posted Items:** Facebook Share is a Facebook application that lets you promote any Group, Event, Photo, Link, or Application you come across by giving it real estate in your "Posted Items" list on your profile page.

You can also send it directly to your friends' Inbox.

- **Mini Feed and News Feed: Facebook's news feed is like a wind that blows your marketing seed in Facebook.** While you're not able to publish directly to the feeds, Facebook's Mini Feed and News Feed archive your users' engagement with your brand and syndicate it to their friends, networks, and beyond, amplifying the reach of your campaign by orders of magnitude.

2. Facebook Tools for Advertisers: Facebook always offers both integrated and self-serve solutions to reach broader slices of the Facebook audience. But this process needs some bucks to be spent. Depending on your budget, you can get started as an advertiser on Facebook with as little as a few dollars for a short-run flyer or as much as several hundred thousand dollars for a customized "sponsored group" destination inside Facebook.

Some Tools Related to this Field are:

- **Ads:** Facebook offers advertisers several different types of ads. The type you choose will depend on the goal you have for the ad.
    - Brand awareness image ads
    - Local awareness ads
    - Video ads
    - Link click ads
    - Multi-product carousel ads

    - Offer claims
    - Dynamic product ads
    - Canvas ads
    - Collection ads
- **Polls:** Polls offer an easy way for marketers to quickly conduct research within their targeted audience.

3. Tools for Application Developers: For marketers who can harness technical resources, the Facebook Platform offers the most powerful way to create engaging connections with your target audience on Facebook. There are literally thousands of good apps to build business brand on Facebook. Some of the tools are:

- **Profile Box: The most common way new users** find applications is through application profile boxes on their friends' profile pages. The challenge of profile box design makes it compelling for an existing user to keep it on his/her profile, and appealing enough to a new user to click on and install the application.

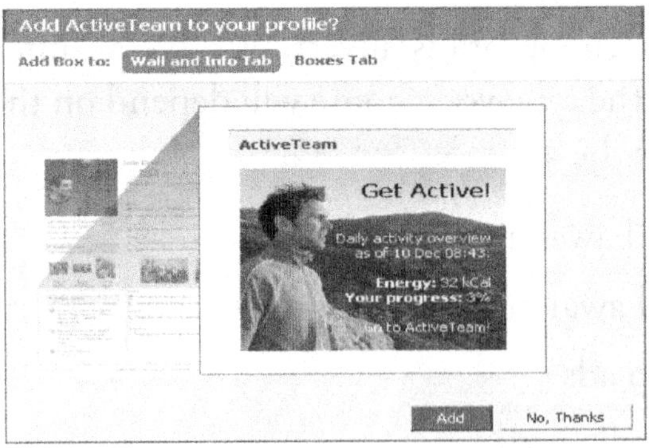

- **Invitations:** One of the most powerful viral channels available to Facebook Platform application developers is invitations. The invitations API allow users of your application to invite up to 20 of their friends per day to install your app.

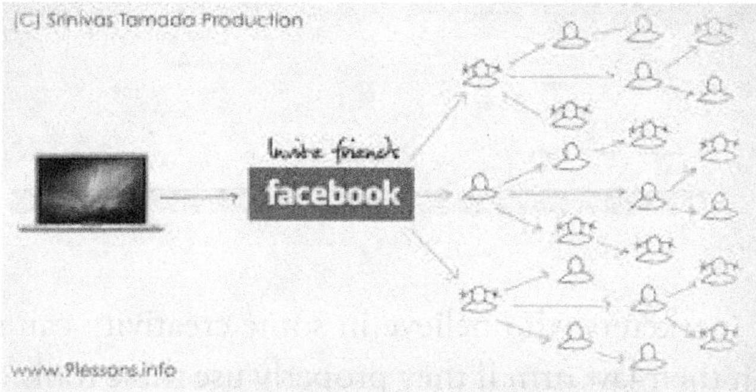

- **Facebook Notifications:** Notifications have been proven to be an effective tool for retaining existing users of your app. However, notifications get less press than feed items and invitations because they're not as effective at spreading your app.

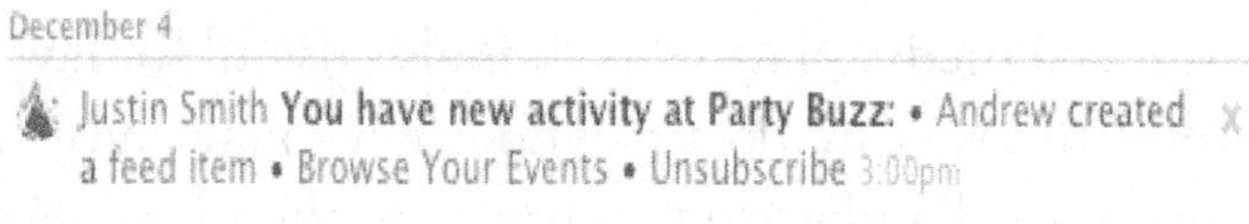

Application Directory: A surprising number of application installations come directly from the Application Directory. It is a great tool to create your brand on Facebook.

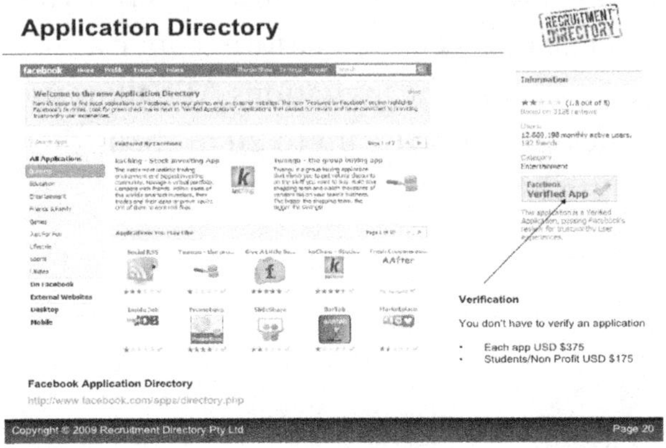

Direct marketers who believe in some creativity can surely shine in their law firm if they properly use these tools. None of the other ways could be better that these tools to build your own brand market on Facebook.

How to Convert Your Facebook Traffic into Sales

There is no doubt that today Facebook has become the leader in social media platforms. It is also useful for law firm owners and Internet marketers who use it to take advantage of this social media platform to promote their business.

Facebook helps them to create their own business fan page, build their brand name and also get traffic. Millions of people are today using Facebook, so it is a great marketing tool for you.

However, converting your fan page traffic into potential sales takes time. There are various ways to use your Facebook Fan Page to generate more traffic into sales for your law firm.

Here are 10 effective tips to use Facebook Fan Page effectively to get sales:

1. **Offer Ultimate Communication with Your Audience**

   Facebook people are very active; they will always check their Facebook accounts often just to update their status. After creating your fan page if you don't provide the same active communication with your audience, then you'll lose their attention very quickly.

   You need to update your page often just to remind your audience that you are there and active. You should also answer the questions promptly if your audience asks something. This builds trust between your business and your potential customers. The more you can earn credibility with your customers the more you will get sales.

2. **Advertise Your Fan Page with Facebook Ads**

   If you already built your fan page you can create buzz regarding your business on Facebook. You can get your page noticed by investing a small amount of money to promote your fan page. Without properly promoting your fan page you likely won't get noticed.

   When you promote your page using Facebook Ads, it won't sound like you're pitching Facebook users with a blatant sales page. If you use your fan page as your landing page in your Facebook Ads, then more new readers will come to your fan page because they perceive your ad not as a sales page, but as an interesting page that they want to explore. Automatically your traffic will convert into sales.

3. **Provide Useful Information, Not Just Product Promotion**

   You should also balance your fan page with useful information related to your niche. Using your fan page as your prime promotional tool to promote your product will be less effective because most Facebook users aren't ready to buy anything yet.

   They came to Facebook for fun, social interaction, and information exchange. They'll love your fan page if you provide them with good information. Instead of solely promotional posts, use your fan page to educate your customers. When they find interest in your information, that will lead to interest in your law firm which will lead to sales.

4. **Don't Hide Any Information**

   Many fan pages don't show their original company names to their fans. They even don't list their postal address. If your fans can't see your address or know exactly who you are, they will be less inclined to do any kind of business deal with you.

   If you are serious about your law firm, show your visitors that you have nothing to hide. List your full contact information on your page.

5. **Show Your Visitors You Are Different**

   Show your website visitors how you're different and why they should use your products or services instead of your competitors.

Suppose you provide Personal Injury services. You might say something like: "We distinguish ourselves from other law firms by offering boutique service, where every client works directly with a lawyer and not a case manager."

6. **Adding a Review Tab to Your Fan Page**

   Not every Facebook Fan Page has a review tab. By default, certain Fan Pages have the review tab in place.

   A review tab is like having a verified testimonial. In order to provide you with a review the user must already be on Facebook, have an account, and cannot be associated with your page as an administrator. Having these reviews provides prospects with a sense of security knowing that others have already used your company's products or services.

7. **Promote on Other Sites**

   You should put your Facebook fan page button on your company blog, websites, and on other social network sites. The more sites that have your links the more people have an opportunity to sign up on your fan page. By getting more visitors you will get more chances to get sales.

8. **Maintain Uniqueness**

   Facebook has 2.45 billion users. It's not possible for all fan pages to attract all users one at a time. If you want to stand apart from the competition you should maintain a unique fan page to your niche.

   Look at related pages in your niche just to see what

others are doing. Take notes and think about how you can make your fan page better, because this will help build brand recognition.

9. **Create Competitions**

   Another idea to get sales from Facebook is to arrange for some attractive competitions and giveaways. People like to receive free things, and if you offer something of real value, it will create tremendous buzz.

   Practicing this continuously will lead to more interest from your traffic which will lead to more sales.

   Check with your state on what regulations there may be around this.

10. **Add a Contact Form on Your Fan Page**

    Facebook fan pages don't provide an easy way to contact you. This means you need to create a contact form on your fan page to easily allow a visitor to reach your business.

    If the visitor is genuine, this contact form will bring them to you. But if they see that there is no way to contact you quickly, they may lose interest in your law firm.

These marketing strategies should allow you to generate more sales for your law firm. Taking time to implement these strategies will provide you with many more opportunities to reach your target client.

TEN

Video Marketing –

How you can tap into the POWER of YouTube AND other video sharing websites to enhance your visibility and drive better conversion

Did you know that YouTube ranks second in search engines? Yes, it's actually ahead of Bing & Yahoo!

Most law firms are extremely focused on search engine optimization but neglect the opportunities that video and YouTube provide. Implementing a video marketing strategy for your law firm can get you additional placement in the search results for your targeted keywords, enhance the effectiveness of your SEO efforts and improve visitor conversion.

Why Use Video Marketing?

There are a number of reasons to use Video Marketing for your law firm.

First, it will increase your exposure on the search engines, giving you more placeholders for the keywords that are most important to you. Video marketing will enhance your SEO effort by driving visitors to your website and creating relevant links to your website, which will improve conversion.

Once somebody gets to your website, if there is good video on the home page and the subpages, it will resonate deeper with your potential clients than a site without video. This helps convert those visitors from just browsing around pages to actually picking up the phone and calling your office.

Again, YouTube is the second most used search engine there is. Obviously, Google is number one. One would think that Bing and Yahoo would be the other major search engines, but that just isn't the case.

There are significantly less videos than there are web pages on the Internet. So creating relevant and quality video content for YouTube and other video sharing sites is a really huge opportunity.

These videos will help you to connect with people and answer their questions when they're looking for information on what you do.

I talked about the fact that you can show up in search engines with an image next to it, and you can obtain multiple placeholders on Google for the keywords that are most important to you. Here is a screenshot representing what this means:

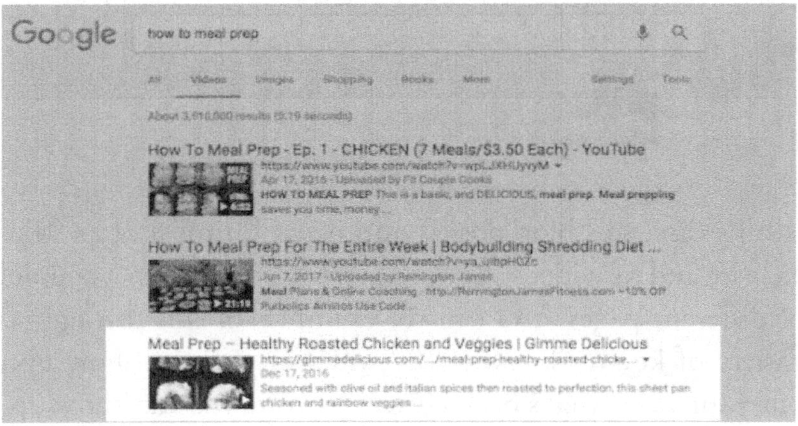

This website, gimmedelicious.com, really likes to focus on meal preparation. This particular blog explains the basics of meal prep. If we type in a search on Google for "How to meal prep," you'll see that his video shows up on page one with a few other listings. If you click that link, it takes you straight to a video.

If you do this right and you optimize your videos correctly (I'm going to show you exactly how in this chapter), you can start to have your video show up in the natural search on Google, which is extremely powerful. It also gives you the opportunity to have more placeholders for the various services that you provide.

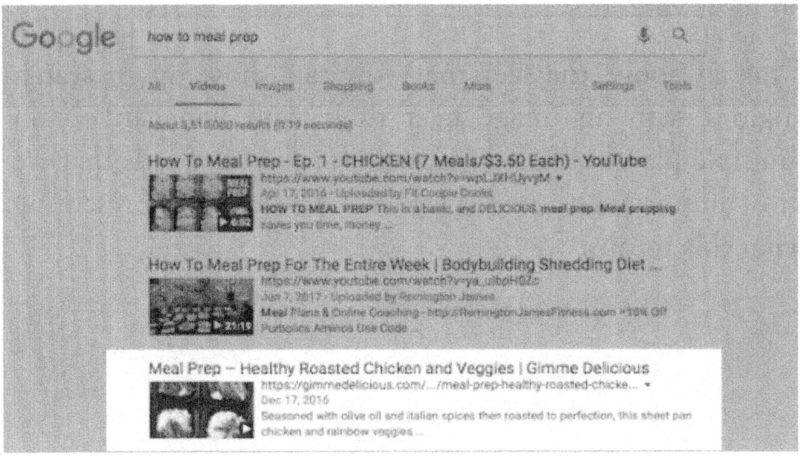

In this case, this blog post is optimized for meal prep. The website and his videos show up in organic search in a couple of different pages. You can see examples of this through a variety of keywords. Not only will your website show up, but your video that's been correctly optimized can show up in that search as well.

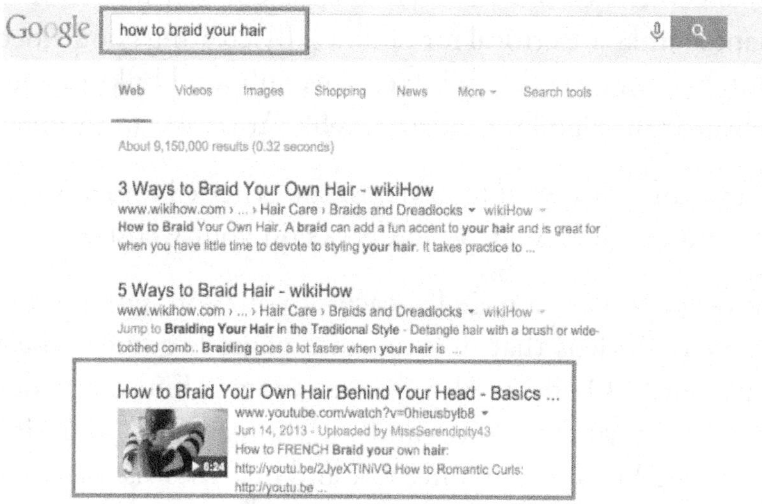

Another example is this YouTube video that describes hair braiding. If you type in "How to braid your hair," you will find that the Wikihow website ranks first and second, followed directly by a video.

Whether your video ranks number one, number two, or number three, most of the time you'll notice that the video gets a much higher clickthrough rate.

This is because there is a large image displaying what that video is. Where are you going to click when you're searching? A static piece of text or the image? A lot of attention goes to those video results. This chapter explains how to incorporate these types of videos for your products and services and optimize them to show up just like this in your market.

Video Helps with Your Overall SEO Effort

The other thing that we can accomplish with video is the

enhancement of our SEO efforts. As covered in the SEO chapter, links are critical for ranking. By creating good video content, you have the ability to drive inbound links to your website from high level video sites like YouTube and Vimeo.

Again, you don't want to have just the generic Home, About Us, Our Services, Contact Us pages on your website.

You want to have a page for each of your core services and products. Videos that link to those pages is going to help with that SEO effort. Also, you're going to find that video content on your website, and on the pages of your site, actually reduces your bounce rate and increases visitors' time on your site.

These are SEO factors. 'Bounce rate' refers to somebody getting to your page and clicking back immediately or browsing away. Google understands those actions as the page not being relevant to that search.

If the majority of the people that get to your site click off and leave right away, your bounce rate is high, and Google is going to start to show you less prominently in their results. That's part of the Google algorithm. The other factor is the amount of time spent on the site. If somebody gets to your page, stays there for ten seconds, and then moves on, the visit might not get treated like a bounce, but Google is still looking at the length on the site.

If you have a video and a visitor takes the time to watch it in its entirety, that's improving your website visit length statistics. Even if they only watch a couple seconds of the video, you have captured their attention long enough that

Google is going to see your site is relevant.

Don't get confused by the notion that having video on your page automatically improves your SEO. That's not necessarily the case. But having people stay on your page longer and not bounce off does impact SEO. Here's an example of how you can drive some links with your videos.

Here's an example of the YouTube channel on "Personal Injury law firm", which is a professional channel from an Atlanta Personal Injury law firm client. You can see in the description area that they have used their key search terms.

This link now connects you from YouTube to their site, and that's a relevant high impact link. You can do the same thing. I would suggest that you do not link to the home page, but rather to the specific subpage related to the video, like the interior painting tools or the surface preparation page.

Here's an example of the personal injury website.

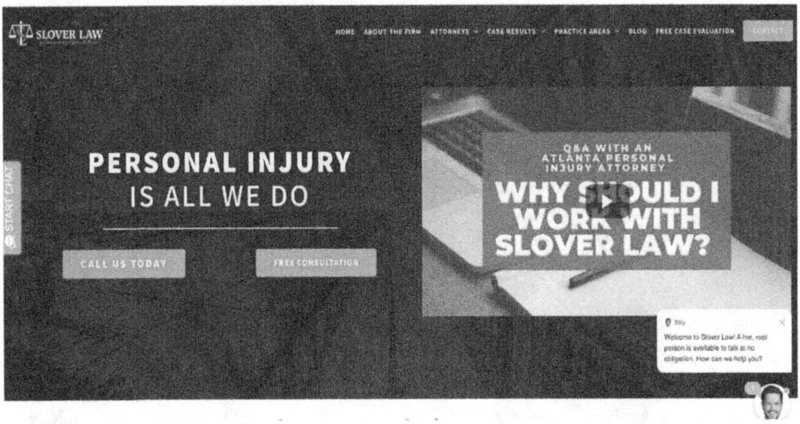

If you check the website, you will see a nice video above the fold section of the website. If you are interested in the service offering, you're most likely going to watch that video on the home page. Again, this is how you improve visitor time on your site and decrease the bounce rate.

People like to watch videos. It's very rare that you're going to see a video on subpages, but you'll find that if you do have that video content on the homepage banner or on your services page, people will take a couple minutes to watch it. Video is unexpected and it's more interesting than text. People enjoy watching someone explain the topic that they are researching.

Above the Fold

Keeping a video above the fold means that you don't have to scroll down to see the most important information. Notice that as soon as this page is pulled up you immediately see the video.

You will want to do the same, providing an intro video about who you are and what you do. That video, placed correctly above the fold, will improve on-page site time and reduce your bounce rate.

I've mentioned the fact that video gives you more placement in search. It's going to give you better search engine optimization because you get the links from the video sites, you're improving your time on site, and reducing your bounce rate.

But the benefit of video, that is probably even more powerful than anything, is that it's going to improve conversion.

You can have the best SEO strategy in the world and drive hundreds of people that are looking for your services to your home page or to your subpage daily. But, if it's not converting and people aren't picking up the phone and calling to hire you for your services after they visit your site, you're missing a major opportunity.

Improving Conversion with Video

Improving conversion is one of the main things having intelligent video on your site will do for you. The fact is that video clips resonate with people. They like video because it gives them the chance to get to know and trust you before they call you, especially if you follow my strategy rather than creating a super corporate video.

If you create authentic video of your team, the owner or your paralegal talking directly to the camera, connecting with you on an emotional level, answering questions and giving a strong call to action, your conversion rate will improve.

Video also gives you the ability to connect with different modalities. Everybody thinks in a different way. Some people are readers and will read all the content on a page. Some people are listeners, so if there's the opportunity to listen to something rather than read, they'll choose to listen.

Other people like something visual. Motion grabs their eye. By having video on your website, combined with text (I'm not saying to abandon text), you have the opportunity to connect with every type of person. Some people will watch the video and only connect with that, because they wouldn't take the time to read a plain text web page.

Here's a website of an online institute that offers private investigation training.

Upon visiting their website, you will see both the text and a video that explains what PSI has to offer. If you watch the video, and if the lady says something that's relevant to you that makes you feel connected, you may think, "Man, this is the right institute that I was looking for."

Do you think you would be more apt to call him over somebody who doesn't have that type of messaging on their website? Video on the site is going to improve conversion rates. It's going to make your phone ring more.

Leveraging Video

How can we leverage video? We understand that it's powerful, it's going to improve your SEO, it's going to help us get better placement on the search engines, and it could potentially help with conversion. How can we expand upon this?

What we want to do is create simple videos about your law firm, your services, and the most frequently asked questions. You are then going to upload those videos to YouTube and other videosharing sites, and syndicate them to your website and social media profiles.

What type of video should you create? Like I keep saying, "People resonate with people." Keep it simple, be real and be personable. Put your real face on the camera, or the face of someone that represents your company. Be frank and to the point. It doesn't have to be a 20-minute video. An appropriate length would be 30 seconds to three minutes long, enough to get the message across.

Don't Over Think It!

Don't feel like you have to go all out and hire a highend production crew or go out and buy an HD camera in order to make this happen. The reality is, you can create video clips using technology that you already have. If you've got a 4g Smartphone or a webcam, you have the ability to create video content that will work for your website. Today's iphones are amazing and can create what we are looking for here. Don't get 'analysis paralysis'! Just take action!

You don't need highend editing software either. YouTube gives you the ability to upload regular video and edit it right within the system. By edit, I mean cropping and tailoring the video to begin and end where you wish. You can put your phone number down in the bottom area of the video as well as a link to your website. Or you can use simple editing software like iMovie (free with Macintosh computers) and Movie Maker (free with the PC).

Using the technology that you already have, stand in front of a company sign with your logo or in your office, walking in front of a courthouse our police station, and talk to the camera; talk with the people that are visiting your website, because that's going to stick with them.

What Kind of Videos Should You Create?

You can create just about anything you want. But the ones that are going to be most relevant are the ones that pertain to your services. Your clients have fears, and if you can address the top priorities they have you will be adding value and engagement.

The first video that I recommend you make is an introduction for your website. This can be as simple as, "Thank you so much for visiting the XYZ law firm website. We specialize in providing XYZ services to the XYZ area. These are the things that make us unique and why people tend to choose us. We'd love the opportunity to serve you. Give us a call right away at the number below, and we can send somebody to your house to resolve your XYZ issue right away."

A simple video along those lines should be the first step of your plan. It's a necessity.

The other videos that you want to create should be about your primary services. This ties in well with the SEO strategy discussed previously. You want to make sure you have a page on your website for each one of the services that you provide.

If you are a Divorce Lawyer, you don't simply provide general Family Law services. You offer mediation, child custody, adoption, property division and more. Make a list of the services that you want to attract more business for and shoot a brief video about each.

The other very powerful piece of content that you should incorporate, but should be phase two, would be your frequently asked questions, or FAQ. Make a list of the questions that people tend to ask and create a little video about it.

A few examples include: What to do if I'm served divorce papers; how is child support determined in divorce; what if I can't afford an attorney for my divorce. Get creative!

This is common information to you, but the average client doesn't know. Creating a little video providing answers to these frequently asked questions makes for great video content for your YouTube channel, to be syndicated on your social media profiles, and/or uploaded to your blog on your website.

Sharing Your Message

Now that we know what types of videos we want to create and how to create them, what should we say? Should we have a script? Should we wing it?

You want to be natural, you want to be authentic, and you want to be real. Some people must have a script because they don't feel comfortable doing video outside of a scripted methodology.

But, if there is any way you can get in front of a camera and speak naturally like you would to a customer in person about your services, that's going to work best.

What to Say

Here is a simple script you can follow: "At XYZ Company, we provide a full range of XYZ services (to the specific area, whatever area you're in, or whatever service this video is about)." Have a brief description of what you do in that area, and then, "If you're in need of this service in your area, we can help. Call our office today at 555-5555."

A simple video like that for each one of your services should always include a call to action telling them what to do. Also, if you feel comfortable with it, referencing a discount could go a long way.

Don't over think this. Think about the core services that you offer. Shoot a quick 30-second to one-and-a-half-minute video about each and you're ready to roll.

What to Do with Your Video Content

What are you going to do with the videos once you've got them? Now that you have completed shooting your videos, what you want to do is setup a YouTube channel.

You can do this by going to YouTube.com. You want to upload your video, name it correctly and intelligently, putting it in terms that people will use when they're searching. If somebody is looking for DUI services, they are going to type in "your city DUI Attorney." You want to name the video using your keywords.

When you upload it to YouTube, you want to title it "Atlanta Divorce Lawyer" or "Miami Bankruptcy law firm" and then put a description with a link to your site. "Visit us online at yourlawfirm.com/chapter13" and then include a description about what you do, briefly outlining what was said in your video.

YouTube Best Practices

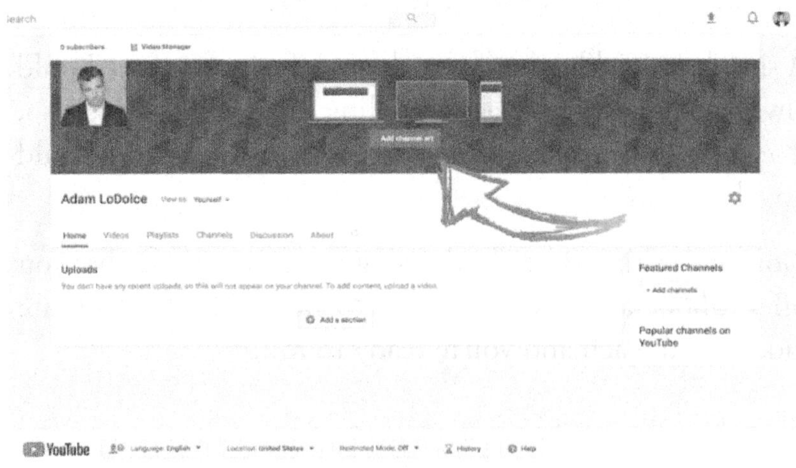

When you setup your channel, make sure that you give it a "city plus service, name of your law firm" title, instead of just your law firm name. You are also going to add tags with keywords to it. Don't just leave the tag area blank.

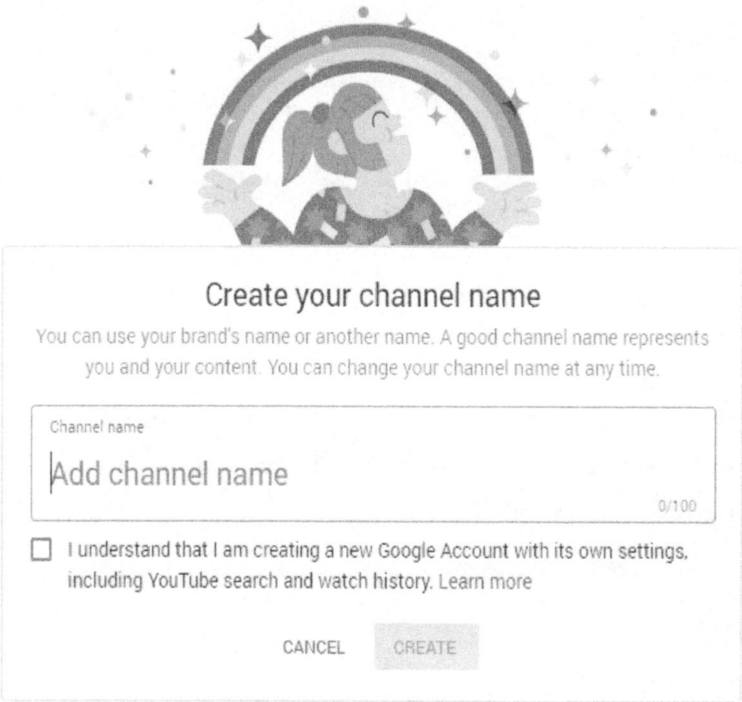

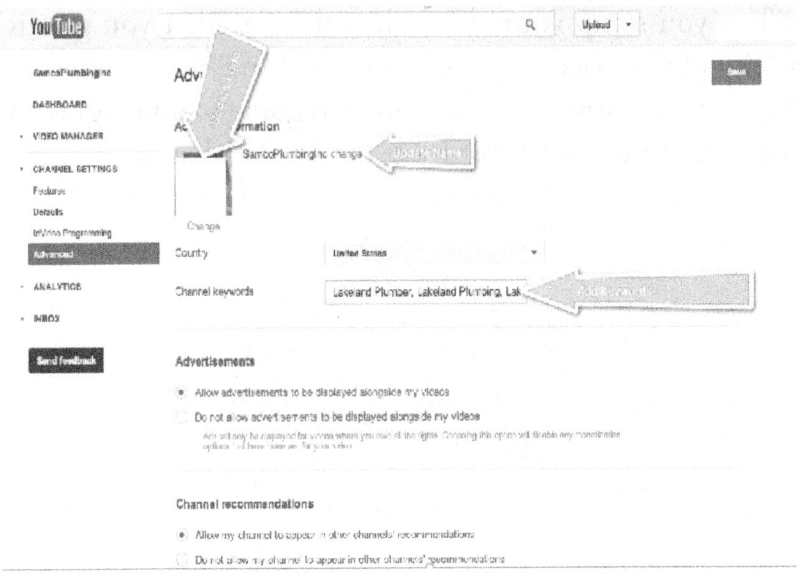

Make sure you use your name, address and phone number in every description on your YouTube channel because this is a good citation source.

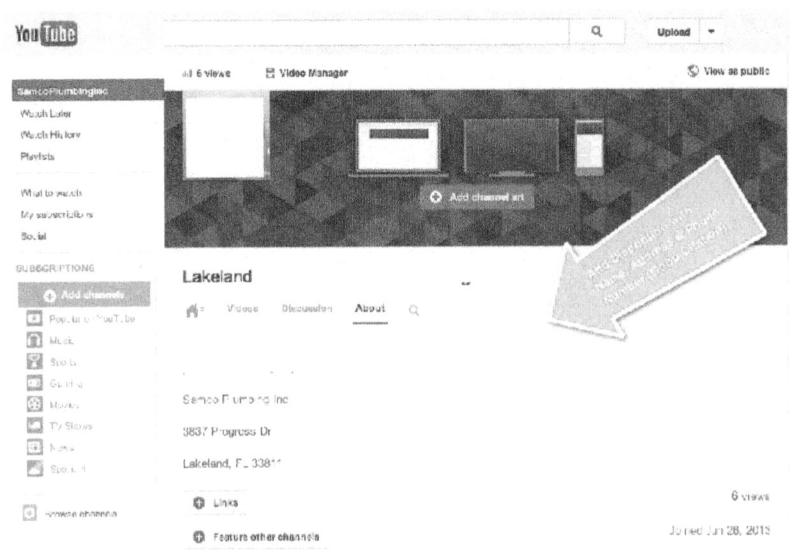

As covered in the Google Maps optimization chapter, citation development is critical (having your law firm name, address and phone number referenced consistently across the web). This is a great place to get citations. Also, make sure that there's an image avatar with your law firm logo. You can update the default image by putting in your logo or put a picture of the team or office.

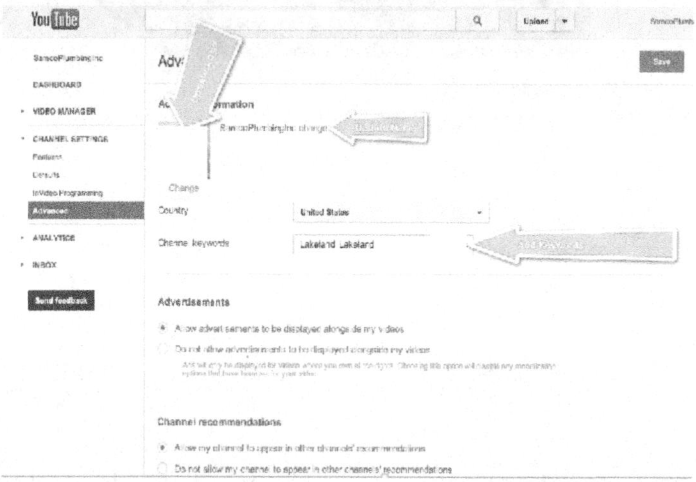

Here's some visual representation of this. If you log into YouTube and create your channel, you'll get an email confirmation. Once you're set up, you can go to the "My Channel" settings and make some of the updates I referenced on the previous slide.

To change your logo, simply click "change" and choose your image, a very simple step.

Where it says "Your company name," it's going to default to something basic such as your email address on Google. You can hit "change" and update it to say "your city family lawyer" or "your city DUI law firm" and then a dash and your law firm name.

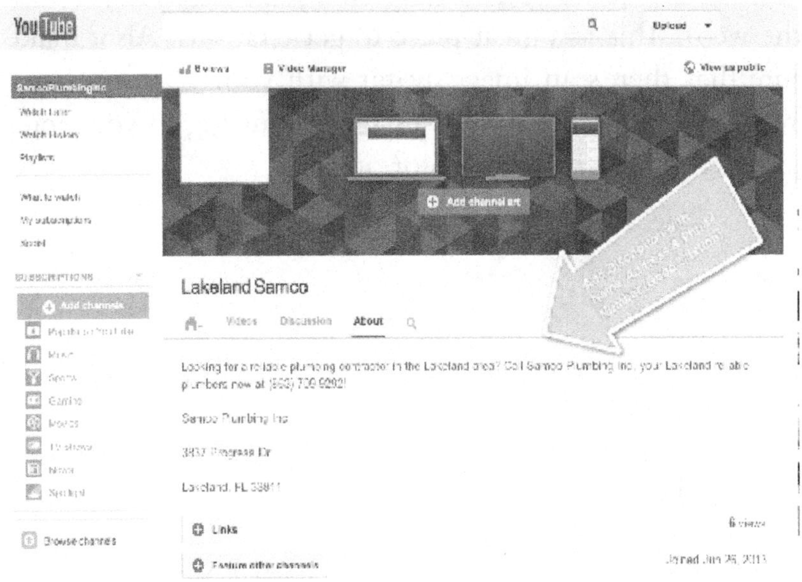

This gives you the chance to get your YouTube channel itself to show up for your keywords in the search engine. You will also have the opportunity to add your channel keywords. That is where you can type in words such as "your city divorce lawyer," "your city estate law firm," "your city family legal services," and of course your law firm name.

From there, there's a section where you can click "About your company" and put a description about who you are, what you do, and what areas you serve. You can get as creative with this area as you want, but it is most important to make sure you first put a description of your services, and your city.

If you're in Tampa, you put Tampa. If you're in Lakeland, you put Lakeland. If you're in Los Angeles, you put Los Angeles. Put your phone number and, again, restate your name, address and phone number. Citations are important. Having this in the description area is a powerful citation source.

Always put your law firm name, address and phone number (NAP) the same way as you did on your Google Map listing, your Angie's List listing, etc. That way, you will be consistent across the web, improving the probability of ranking in the Google Map listings.

Video Tagging Best Practices

Now, let's talk about video tagging best practices. Let's say you created the inventory of videos I recommended: an intro video and clips for each of your services.

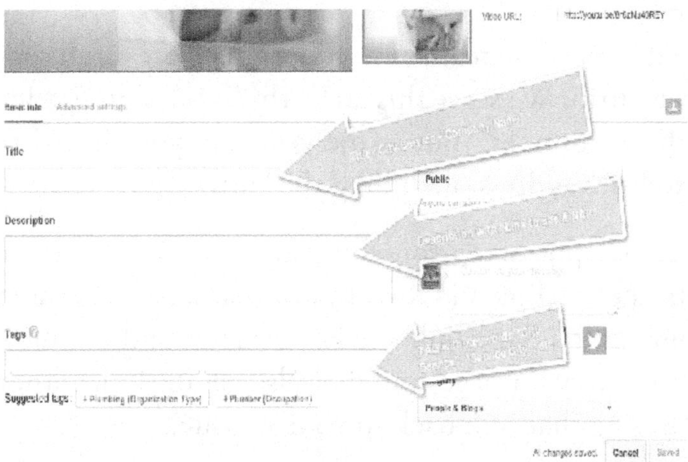

How did you tag those videos to maximize the opportunity and to make sure you're going to rank well in search?

- Title Video with City Service – law firm (always mix this up a little)
- Description should always start with http://url.com and then describe the service using those same keywords. ALWAYS ADD N.A.P. (Name, Address, Phone) INFO AT THE BOTTOM OF THE DESCRIPTION
- Use your keywords as tags and include the law firm name
- Choose most appropriate screenshot
- Click "advanced settings" and add address to video

The first thing you want to do is have your primary keywords in the title of the video as well as a description that includes the "http://" before your web address.

In the description area, you can put in "We're a full service XYZ law firm. We serve this area. This is our name, address and phone number," but at the very top, you should have your website address, including the "http://".

If you just put www.yourcompany.com, YouTube won't understand the link and it will show that it isn't clickable. If you put "http://" the link will be clickable, and visitors will go straight to your page, and they also get the link authority from having that link back to your website.

Choose the screenshot and add video. Whenever you upload

your video you are able to control your title and your description, as well as the ability to add tags.

Titles Matter

Again, don't call your videos "your law firm name." Don't call it "law firm." Don't call it "attorney." Call it "Your city + that service," and then your law firm name. Title your videos the same way that somebody would search.

If it's your intro video, you might want to call it "your city + your primary service" Example: you're a local "Lakeland Divorce Attorney – XYZ law firm." If it's the child custody page, "your city child custody lawyer – XYZ law firm".

It is really critical that you have the right titles on your video. This is what is going to make it so Google can locate it and include it in search results.

The next thing you want to do in your description is to put the link at the very top. The first thing you want to do is include a link back to the home page or to the specific page that you're discussing in the video.

If it's the child custody page, don't put a link to your home page. Put a link to that child custody service/content page, and again "http://yourlawfirm.com" make sure you've got that "http://".

Below, you add your tags. Within those tags you can put in your city lawyer, your city law firm, your city child custody, and everything in between.

What Else Can You Do with Your Videos?

Now that you've updated your video and you've properly optimized it, your title is correct, and your description is posted, how can we use these videos? Where are we going to leverage them? Well, to really get the benefits of that conversion component, we need those videos to be posted on our website and social profiles as well.

The best way to do this is to copy the "embed code" and post the videos right on your site. The intro video should be embedded on the home page and the service-specific videos should be posted on the appropriate subpages. The way we do this is right within our YouTube channel or YouTube account.

Go to the video manager and find the list of all the videos that you have. Choose the video that you want to post on your website and choose the share and embed option.

You will then be provided with this little piece of code, and that's what I have highlighted on the screen. It goes from Iframe to Iframe. This is the specific code for that video.

If you are updating your website on your own, copy and paste the code right into your website's HTML. If you have a detached web manager, send the code off to them with details on where you want it posted.

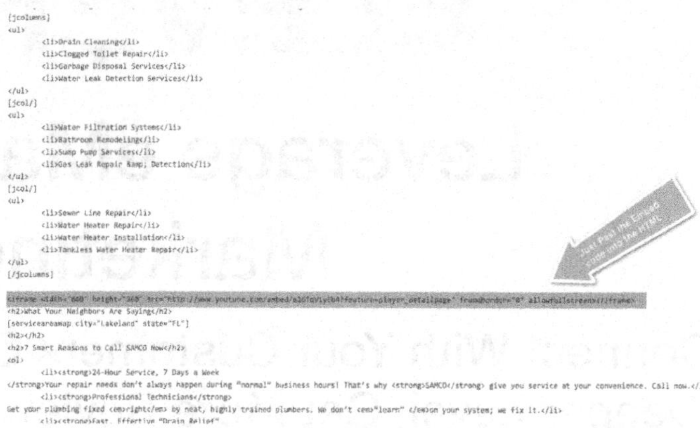


```
[jcolumns]
<ul>
        <li>Drain Cleaning</li>
        <li>Clogged Toilet Repair</li>
        <li>Garbage Disposal Services</li>
        <li>Water Leak Detection Services</li>
</ul>
[jcol/]
<ul>
        <li>Water Filtration Systems</li>
        <li>Bathroom Remodeling</li>
        <li>Sump Pump Services</li>
        <li>Gas Leak Repair &amp; Detection</li>
</ul>
[jcol/]
<ul>
        <li>Sewer Line Repair</li>
        <li>Water Heater Repair</li>
        <li>Water Heater Installation</li>
        <li>Tankless Water Heater Repair</li>
</ul>
[/jcolumns]

<iframe width="640" height="360" src="http://www.youtube.com/embed/a16TqViyLb4?feature=player_detailpage" frameborder="0" allowfullscreen></iframe>
<h2>What Your Neighbors Are Saying</h2>
[serviceareamap city="Lakeland" state="FL"]
<h2></h2>
<h2>7 Smart Reasons to Call SAMCO Now</h2>
<ol>
        <li><strong>24-Hour Service, 7 Days a Week
</strong>Your repair needs don't always happen during "normal" business hours! That's why <strong>SAMCO</strong> give you service at your convenience. Call now.</li>
        <li><strong>Professional Technicians</strong>
Get your plumbing fixed <em>right</em> by neat, highly trained plumbers. We don't <em>"learn" </em>on your system; we fix it.</li>
        <li><strong>Fast, Effective "Drain Relief"
```

Once the code is embedded in your HTML, it will show up on the page itself. That's what we really want to do with these videos. And, of course we don't have to limit ourselves to YouTube. There are a lot of very well known video sharing sites out there.

ELEVEN

Leverage eMail Marketing

To Connect With Your Customers On A Deeper Level, Get More Reviews, More Social Media Followers And Ultimately More Repeat And Referral Business

Ever since there's been email, there's been email marketing. Email marketing is one of the oldest forms of advertising your law firm on the Internet.

Although it gets a bad rap because of all the spam going around, it's still one of the most effective forms of marketing.

I am a big believer in email marketing. It's a powerful way to get instant traffic to your website and getting the telephone to ring, but there is a right way and a wrong way to use it.

Did you know the easiest client to sell to is the client you already have?

Every self-proclaimed marketing expert will tell you that's nothing new. With that said, many law firm owners hardly ever market or keep in touch with their existing client base. Companies will spend thousands of dollars trying to get new customers but never think to market to the clients who already buy from them.

Why is that? I have a lot of ideas about this. I suspect law firm owners think that once a customer buys from them, they will just keep coming back on their own. Or maybe they simply don't want to bother their customers. The truth is customers want to hear from you and they want to be touched by your law firm. If you don't, your competition will.

How Do We Start an Email Marketing Campaign?

The first thing you need is an email marketing service. You shouldn't do this yourself for several reasons:

- Your Internet Service Provider (ISP) will blacklist you for sending bulk mail.
- You would have no stats for tracking your open emails
- It would look unprofessional coming from your Microsoft Outlook box.

With that said, let's take a look at some of the popular email marketing services, all of which are paid services and are priced based on the amount of emails you send. They all start at around $15.00 per month to send a couple hundred emails.

Constant Contact

I have used Constant Contact in the past and I like it for several reasons. It has great tracking stats, the ability to post to your social networks and a relatively user-friendly interface.

Constant Contact has many templates available for use. You can also add your own custom templates. I think custom templates are a MUST for any law firm wanting to promote their brand. You will have to know a bit of HTML but if you don't, you can have a web designer create one for you at a fairly inexpensive cost.

MailChimp

Mailchimp is another service I have personally used and recommend. It's relatively easy to use and offers similar features to Constant Contact. The interface is clean and easy to use. Prices start at $10.00.

iContact

Personally, I have never used iContact, but after reading about it on their website www.icontact.com, it looks fairly intuitive and similar to both MailChimp and Constant Contact.

I think all of these services are a good solution for any law firm looking to add email marketing to its Internet marketing strategy.

How to Get Email Addresses

I am asked on a regular basis about how to get email addresses. It's not as easy as sending a letter in the USPS mail to anyone you want to. The reality of it is that just because they are your client and you have their email address doesn't mean you can send them anything if you don't have their permission.

This certainly is a fine line, because you somehow already have their email address, and they have used your services before, so is it really considered spam? Technically, yes. You didn't ask them if you could send them specials or a newsletter in email form.

The first thing you really want to do is get your clients' permission to add them to your email list. There are a variety of ways to do this, including placing a form on your website, putting a sign-up sheet on your counter or even a putting a space on your job ticket that they sign when you complete your service.

Explain that you send out tips about your industry or specials on a monthly basis and would love to have them on your mailing list. You might even offer a discount coupon off your services if they sign up.

Getting that email address is valuable, so if it cost you 5%, go for it. Remember, you want the opportunity to have your law firm's name in front of your customers every single month. You want to remain top-of-mind if one of their friends is looking for services like yours or if they run into an emergency.

I had a pest control service provider come to my home several years ago. He did a good job and was very professional.

Four or five years later, I needed the services of the company again. I lost his business card and could not remember the name of the company. I had to find another pest removal service. He lost the business because he never stayed in contact with me. It was a big job that he lost, $1,500.00 to be exact.

Start building your list today.

What to Send and How Often

First, what do I send? You must use the 80/20 rule, 80 percent good information and 20 percent sales. If all you send is emails about what services you offer, no one will ever read it. It's a great way to kill your list.

Draft up some information about your specialty, give good tips, and make sure it's information that will help your users.

How often you send your emails is very important. I always go with once per month, around the same time every month.

It is important to commit to a date. More than once a month is too much and annoys people.

I get an email from a company I purchased from in the past and get 3-4 emails a week from them, 100% sales, sometimes several times a day. I HATE IT and it drives me nuts. I removed myself from that list very quickly as I'm sure others have as well.

Get Legal

Make sure you have allowed customers the option to Opt Out of receiving email messages at the bottom of every message.

Make sure that it's easy because nothing is more annoying than receiving emails that you don't want. If someone does not want to receive your messages, then remove them from your list.

They may be getting emails from too many sources and just want to clean out their email box. It does not mean they will never buy from you again. But I will tell you this, if they want out and you keep sending email to them, it's a sure-fire way to bother them and they will likely never buy from you again.

Again, you want to leverage email marketing as part of your overall Internet marketing strategy. The best way to use it is to be sure you're collecting the email address from all your clients and prospects.

From there, use email marketing to get online reviews, engagement on your social media accounts and remain top-of-mind as a strategy to get more repeat and referral business.

What is the Best Time to Send an Email Campaign?

These general email send time tips are widely accepted by the email marketing community. They are great when you're starting off, but be sure to read on and see why they won't always work.

Day-time vs. Night-time. While this one may be obvious, it's usually better to send out your email campaigns during the daytime. You know, when people are awake. Not asleep.

Mad Mondays. The general consensus is that you should avoid sending out email blasts on Mondays. Why? People are already bummed out about the end of the weekend. They march into the office and are flooded with emails they've collected over the past few days. What's the first thing they do? Delete those emails of course!

Weekends. Historically, weekends are the days when folks are out running errands and going on adventures. Weekends tend to have low open rates, so most marketers avoid them like the plague.

Fan Favorites: Tuesday, Wednesday, and Thursday. Tuesday, Wednesday, and Thursday have traditionally been favorite days to send email campaigns, as email marketers seek to avoid the Monday angst and Friday's itchy feet. MailChimp confirms that Tuesday and Thursday are the two most popular days to send email newsletters.

TWELVE

Overview of Paid Online Advertising Opportunities

If we revisit the Online Marketing Plan referenced in chapter one of this book, you will recall that the foundation of your

Internet marketing plan should be focused on the organic, non-paid marketing efforts (Website, SEO, Google Maps, Social Media Marketing, Video Marketing, etc.).

Once you have a strong foundation, you should have the financial resources to invest in other paid online marketing initiatives.

In this next chapter, I want to quickly recap the paid online marketing options you should consider:

- Pay-Per-Click Marketing on Google AdWords and Microsoft Search (Yahoo & Bing)
- Paid online directory listings on sites like MerchantLocal, Yelp.com, YP.com, BBB
- Pay-Per-Lead and Lead Aggregators like Emfluence.com, Fuellead.com, Intellibright.com, etc.

Now, let's talk about the most powerful of these strategies – Pay-Per-Click Marketing.

THIRTEEN

Pay-Per-Click Marketing

(Google AdWords AND Bing Search) - How To Maximize the Profitability of Your Pay-Per-Click Marketing Efforts

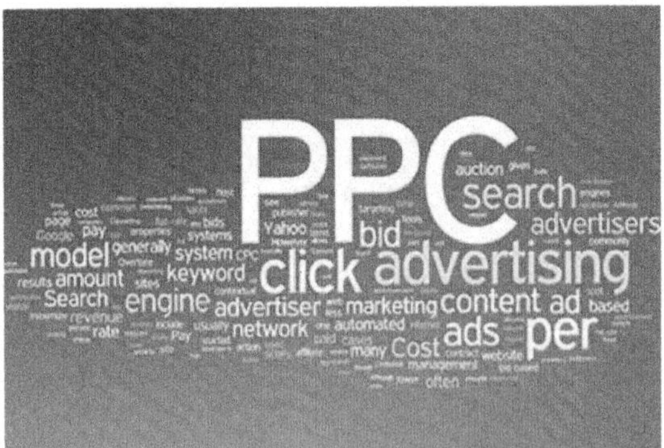

In this chapter, we're going to talk about Pay Per Click

Marketing to help you understand how it works, why it should be integrated into your overall strategy, and how you can run a really effective program that can drive a nice, profitable business for you and your law firm.

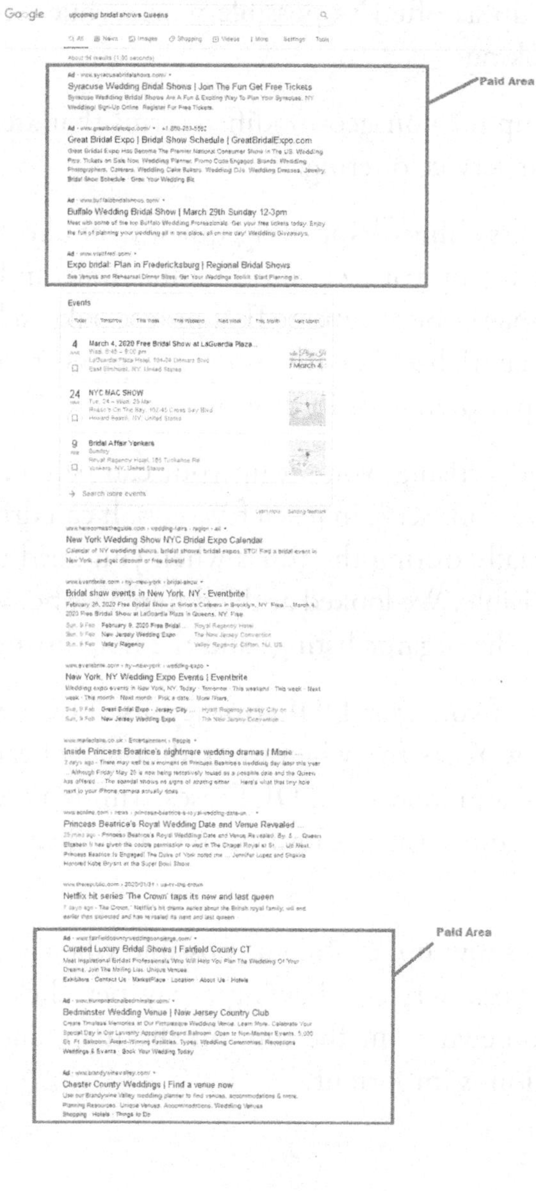

Why PPC Should Be Part of Your Overall Online Marketing Strategy

- Start showing up quickly
- Show up as often as possible where your customers are looking
- Show up for non-geo-modified terms that are related to your service offering.

First, PPC gets things happening quickly, unlike an SEO program, setting up your website, building links and having the right on-page optimization. That process takes a little bit of time to materialize. What you do today and tomorrow, will start to pay dividends in three to four months.

With PPC advertising, you set up your campaign and will start to see your ads serve in just a few days. It can drive good traffic, especially during the times when you need to make sure you're visible. We looked at the differences between the paid listings, the organic listings and the map listings.

For example, if you are a DUI attorney, use PPC a week or so before New Year's Eve when we traditionally head into a higher risk of and increase in DUI cases. This is a great time to advertise your firm because people will likely research that.

You want to show up as often as possible when someone's looking for your services. Having a pay per click ad that shows up somewhere in the top, on the map, and in the organic section is important.

Now you've got the opportunity to show up in multiple places and significantly improve the chances of getting your ad clicked on, as opposed to your competition. A pay per click campaign gives you that additional placeholder on the search engines on page one.

It also gives you the opportunity to show up for words that you're not going to show up for in your organic SEO efforts. This is what I like to call non geo-modified keywords.

SEO and our whole organic strategy give us the ability to show up in search engines when someone types in your city service, for e.g. your city lawyer, your city DUI lawyer, your city attorney, etc. All of those include some kind of geo modifier (your city). They're going to put their city or their sub-city in that search for you to rank.

With a PPC campaign, you can show up for the non-geo-modified terms (Example: family law, custody issues, chapter 13, bankruptcy, estate planning, DUI, etc.), and put in the settings that you only want to show up for people within a 25-mile radius of your office.

If you're in Miami and somebody searches within that area for "divorce lawyer" or "auto accident lawyer," you can set it so that it only shows your ad for the people that are searching within that area. And Google can manage that through IP addresses by isolating where the search took place.

Google can also isolate who ran that search, where they ran that search from, and then place the ads based on the advertisers that are set up for that area.

You only pay on a per click basis, but you're able to show up for those keywords in those major markets. Another reason that you want to consider running a pay per click campaign is because you can run mobile PPC campaigns.

With mobile PPC campaigns, when somebody is searching for your services from a mobile device, it's typically because they need immediate service. They're not as apt to browse multiple pages or listings. Now, if somebody runs a search on their mobile device, and you have a pay per click campaign set up, that search will be PPC enabled.

They can simply hit your ad and automatically be calling your company, rather than browsing to your website and researching.

On a pay per click campaign through mobile, you're actually paying per call as opposed to paying per lead. It's very powerful, and these are the reasons you want to have pay per click as part of your overall Internet marketing plan.

The Pay-Per-Click Networks

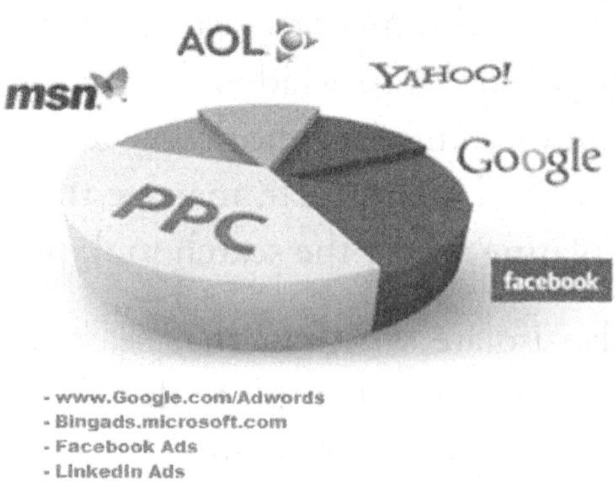

So, what are the pay per click networks? There are two major networks that manage pay per click advertising across almost all the major search engines. There's Google AdWords, which is Google's pay per click program, and then there is Bing, which is through Microsoft Search.

These both have their own network behind them, so when you pay for an ad or pay per click campaign on Google's search network, you're gaining access to AOL, AT&T, USA Today, and Ask.com.

When you get on the Microsoft Bing search network, you're getting access to Yahoo!, Facebook, etc. There are a variety of reasons to consider a Bing Microsoft pay per click strategy.

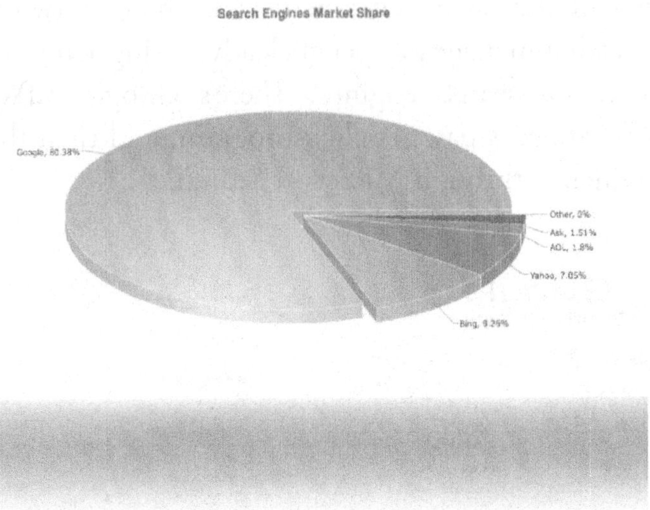

You can review the chart above to see where most people search and what's going to give you the most attention. It clearly shows that Google is the dominant player with no serious competition.

More than 80 percent of all searches happen on Google.com. So, if you had to choose, you would obviously you want to use Google. However, you do get an additional 20 percent by tapping into Bing and Yahoo!.

There are different networks but those two make up the majority of the search market. Running a pay per click campaign on both Google AdWords and Microsoft Bing search will allow you to show up in the majority of the search engines that somebody might be using.

Understanding the Google AdWords Auction Process

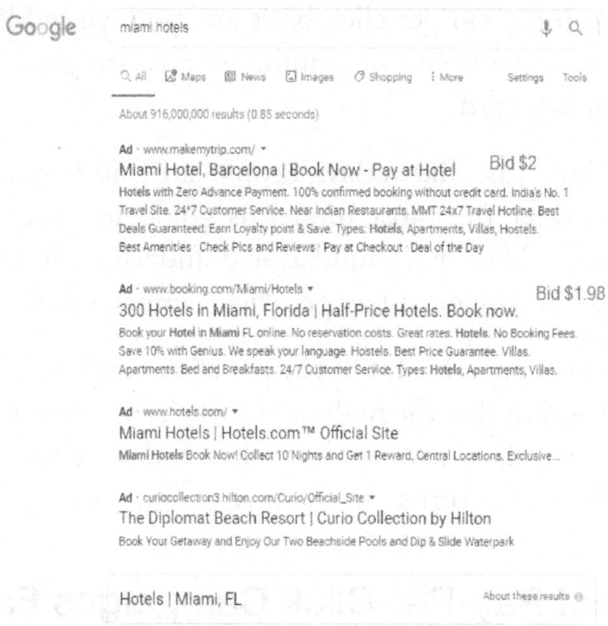

Let's review how Google AdWords works.

In the simplest sense, you're paying on a per click basis and you can choose your keywords (Example: lawyer, your city lawyer, your city DUI lawyer, family lawyer). As you pick those words, you bid, and you pay on a per click basis.

So, let's just say you're bidding on the keywords "San Antonio lawyer," and there are a lot of other lawyer in that city that want to rank for that keyword.

If you say that you'll pay $2.00/click and your competitor says that they'll pay $5.00/click, they're going to be at the top. Assuming nobody else has placed a higher bid, $2.00

is going to be ranked second and $1.20 is going to follow.

I am about to explain why that isn't 100% of the reality. The fact is that you pay on per click basis and you are bidding against the competitors to determine how you're going to rank on your keyword.

It's an auction, just like eBay. People are bidding and whoever can offer the most money is going to have the strong position. With that foundational understanding, we can now explain why most pay per click campaigns fail.

What tends to happen is a lot of pay per click campaigns are built on the notion that the highest bid wins. So, advertisers pick their keywords, throw up the highest bid per click and hope that everything turns out the way they want it.

Why Most Pay-Per-Click Campaigns Fail

- Setup only ONE ad group for all services
- Don't use specific text ads and landing pages for groups of keywords
- No strong call to action or OFFER on the landing page

You might be thinking, you just told me that PPC is a great way to get noticed, and now you're saying that most campaigns fail! I'm going to explain what people do wrong and then show you what to do right so that your campaign is successful.

Typically for example, a divorce law firm sets up only one ad group for all services, whether it's child custody, divorce, property division, lgbt divorce, mediation, etc. instead of different ad groups for each type of service.

Also, there's no specific text ads and no landing pages for those ad groups and groups of keywords.

What you wind up with is the same landing page and the same text ad, whether your customer typed in "child custody, divorce, property division, LGBT divorce, etc." in the search engine.

Whatever was typed into the search engine was likely very specific, and should match up to a very specific page, but that doesn't happen. It all goes to the home page. With this strategy, not only is your campaign going to convert poorly, but your cost per click is going to be higher. I will explain why later in this chapter.

The other reason why most pay per click campaigns fail

is because there isn't a strong call-to-action on the landing page. So, you were just charged $5.00 or $9.00 to get a potential customer to your website and the page isn't even compelling because it does not have a strong call-to-action. It doesn't tell the consumer what to do next.

If you factor these common reasons that pay per click campaigns tend to fail, you can better prepare yourself and set yourself up for success in the way that you execute your pay per click marketing.

Understanding the AdWords Auction Process

$$\text{ACTUAL CPC} = \frac{\text{AD RANK TO BEAT}}{\text{QUALITY SCORE}} + \$0.01$$

Advertiser	Max. Bid	x	Quality Score	=	(Ad Rank)	Actual CPC
Mary	$2	x	10	=	20 (no. 1)	16/10 + $0.01 = **$1.61**
Tom	$4	x	4	=	16 (no. 2)	

Let's talk about how the AdWords Auction process works. It's not as simple as the highest bidder winning. It's more complicated than that.

The reality is Google needs to feature the most relevant results because their endgame is to get people to keep using their search engine over the competition. This is how they can keep their traffic up.

Google can keep their usage up and maintain that 80 percent market share but can also run AdWords and make billions of dollars per year. Ultimately it all comes down to relevancy.

The second they sacrifice relevancy for dollars, is the second they start to become less of a player in their market. So, Google had to figure out a way to make their pay per click program grow around relevancy. And so that's why they established the quality score. They need to make sure that the person or company who has more relevancy gets a higher quality score and as result, can have a lower cost per click.

The way I like to explain it is, if I go to Google and I type in "BMW," obviously I am looking for a BMW dealer or for information about BMW.

Mercedes could say, "That's our demographic also. If someone types in BMW, they're looking for a high-end vehicle. They are probably in the market to buy. Why don't I bid on the word BMW?" Of course, they can. However, the person that searched BMW isn't looking for Mercedes. So, Mercedes could say, "I'll pay $25.00 for everybody that clicks on me when they search 'BMW'."

But BMW might say, "That's my brand and I am going to compete for it, but I am not going to spend $25.00 for every click on my own brand. I'll pay a dollar for every click." Based on quality score, Google may decide to serve BMW because it's in the best interest of the person researching the brand, the consumer. It's also in the best interest of overall relevancy. That's how quality score works. Quality score is really driven by three core components:

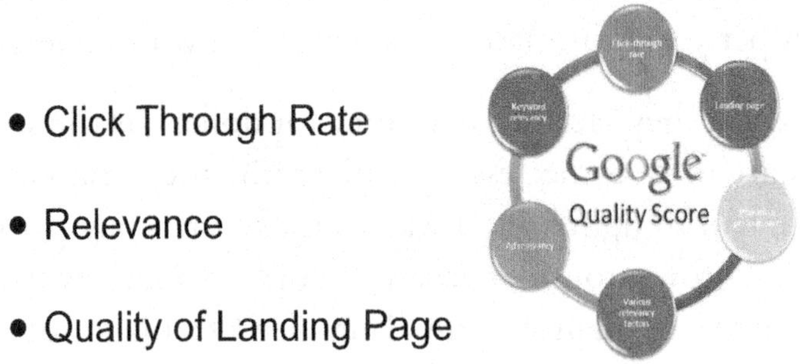

- Click Through Rate
- Relevance
- Quality of Landing page

As somebody conducts a search and your website shows up on the page in the pay per click section, Google is tracking what percentage of those people saw your ad and wound up clicking through.

That's one of the primary metrics that they analyze. So, if your ad is relevant, if it speaks to the person's needs, and if it's compelling enough to them that they click through, Google just made more per click.

This will make them willing to give you a higher quality score because you've got better click-through rate.

Also, relevancy is a major factor. How relevant is your text ad to the keyword that was typed?

Example: If they type in "dental braces," and your text ad reads: "We're the best dentist in the Dallas area," versus "We're the best dentist in the Dallas area and we offer high quality dental braces in the Dallas area."

Which do you think is more relevant to the customer? Google wants their search results to be as applicable as possible. They're looking at your click-through rate, they are looking at the relevancy of your text ad to your keywords, and they are looking at the quality of your landing page.

If your landing page (the page that you drive people to) doesn't match up with what the person just clicked based on your text ad, or if that landing page doesn't have a strong call-to-action and the person quickly returns to the search engine, that signals to Google that you were not very relevant.

This will result in a quality score reduction.

Better Quality Score =
Lower Cost Per Click for Top Positions

By having a higher quality score, you can bid lower and still achieve the top position. This is where you can win in the pay per click marketing game because a better-quality score results in a lower cost-per-click for those who hold the top positions.

Again, if we just look at the reason most pay per click campaigns fail, it's because:

- You only set up one ad group
- You had the opportunity to create a separate ad group for each one of your core services, but you don't use a specific text ad that's going to compel someone to click and improve your click-through rate
- You don't have a strong call-to-action that matches up with what the consumer was looking for
- You're not going to have high click-through rate, relevancy, or an applicable landing page.

All of these issues result in a lower quality score.

You're going to wind up paying more per click. PPC marketing is very competitive. If you're paying more per click, you're not going to be able to spend that much because you won't be getting enough calls to generate return on investment.

The visual representation of this would be like setting up one AdWords campaign for each one of these services (dentist, emergency dentist, family dentistry, cosmetic dentistry, teeth cleaning, dental braces, etc.) and landing people on your home page. That is a recipe for disaster.

That's exactly what you don't want to do.

How to setup your PPC campaign for success

Let's talk about how to position your pay per click campaign for success.

What can you do to ensure the highest probability of success in your pay per click campaign? For starters, set up ad groups based on the specific groups of services that you offer (we're going to map this out using a variety of businesses as an example).

Write compelling text ads that are relevant to your specific keywords or services. Then, link your ads to the specific pages on your site rather than the home page. But, the specific pages on your site that talk about that service should have a strong call-to-action combined with an offer.

What ad groups should you use? What ad groups do you need to set up for your business? Using a dental clinic company as an example:

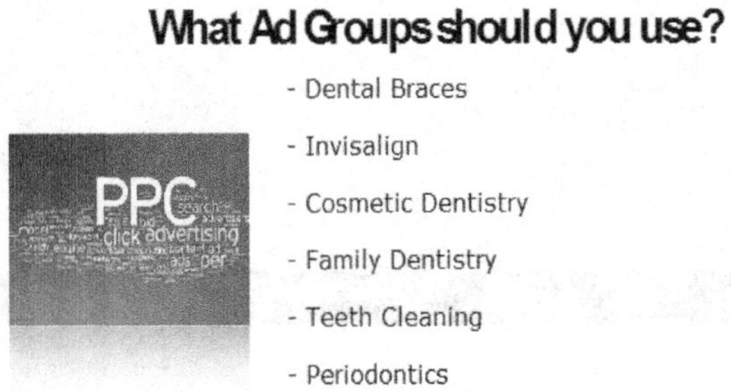

If you are in the dental business, you need to have standard dentistry for the general, "I need a dentist," or "I'm looking for dental services" search. They didn't get very specific. You should have something for that. Have emergency dental services available, for the person who types in "emergency dentist," "emergency dental clinic," "emergency dental services," "24-hour emergency dentist," etc. You want to group those keywords together and have information available for that.

We could go a lot deeper than this, but you should have an idea of what specific types of ad groups you need to set up based on the services you offer. From there, you want to write a specific text ad that speaks to that group of keywords.

Then, you will want to drive them to a landing page on your website that has a compelling call-to-action, that provides what they were looking for and mirrors what your text ad said. I've got a template below to review:

- Pick your list of keywords
- Write a specific text ad that matches up with what those people are looking for
- Drive them to a landing page on your website

Make sure that you've got compelling content on that landing page that emphasizes what they were looking for and prompts them into action, ideally with some type of coupon or special offer, so that they don't move beyond your page and keep looking around.

Wedding showcase AdWords example

Let's look at the bridal showcase example. For an upcoming bridal showcase, in general, you're going to have the following keywords:

- Bridal expo
- Wedding expo
- Bridal show
- Wedding show
- Wedding showcase
- Upcoming bridal showcase

These are the keywords that go into this general bridal showcase ad group. Your text ad should speak to that search.

"Connecting Brides and Grooms with Top Long Island Wedding Professionals. February 10, 11, 12"

You want to pull on the psychological triggers. Are they looking for affordability? Are they looking for quick service? Typically, they are.

Then, drive them to the URL on your site that is specifically targeted at bridal showcase, Yourcompany.com/bridalshowcase. Get them to the page that talks about that specific service.

There are a lot of things you can do on the landing page, but you want to make sure that you tap into that psychological trigger.

MEET TOP LONG ISLAND WEDDING PROFESSIONALS

Free Admission

Brides + Grooms + Family & Friends

Live DJ Showcase at All Shows

Monday, February 10, Marriott Uniondale,

After Hours Entertainment

Tuesday, February 11, Marriott Melville, Variety Music

Wednesday, February 12, Clarion Hotel, Ronkonkoma,

After Hours Entertainment

Live Fashion show hosted by Princess Bridals

Meet with 35 to 40 Wedding Experts at each show featuring Photography, Videography, Wedding Gowns, Caterers and Reception Sites, Limousines, Flowers, DJ's with Musical Performers, Medications, Invitations, Cakes, Tuxedo's, Hair and Makeup Salons, Financial & Insurance Planners, Real Estate & Mortgage Experts, Photo Booth Companies and More! Meet Top Wedding Professionals and get pricing information for your wedding. At Long Island Bridal Expo, Everything for Your Wedding Under One Roof!

Talk about why they should choose you and not the competition and have a link to a page where they can see some external resources.

What does the BBB say about you? What reviews do you

have on Yelp.com? Give them some information so that they can feel confident that you're a credible law firm that's going to follow through on your promise.

Then, have a strong offer with a call-to-action. Get a FREE consultation by referencing the coupon below. Call now! If you have the capabilities built into your website, consider linking them to a form where they can choose to type in their name and phone number and schedule the service right on the spot.

This is a well-crafted ad group specifically for the Wedding DJ keywords.

Wedding DJ

Let's look at Wedding DJ keywords. The consumer typed in wedding DJ, sound DJ, leading wedding DJ, best DJ for wedding, top rated DJs, party DJ, etc. These are people that are in an immediate need of wedding DJ services. You want to group those keywords together, save it as an ad group, and create a text ad that is specific to that.

> *"The Most Tasteful, Talented and Experienced DJs In New England Since 1983. Build Your Event Today. Quick, Free Custom Quotes. Multiple Wedding Services."*

You say just what they want to hear. They've got an emergency and you're entering the conversation that's already going on in their head. You're offering them some type of incentive to do business with you.

Again, you drive them to the Wedding DJ Services page on your website (Example: www.yourcompany.com/

wedding_dj. Then have some content that speaks to that specific situation.

> *"New England's Top-Rated Wedding DJ!!"*
>
> *"Since 1983, we have set new standards in quality and style for wedding DJ entertainment in New England. For over 30 years, we have been the most respected and recognized name in the disc jockey business, offering unmatched service and professionalism to the New England area including Connecticut, Rhode Island, New York, & Massachusetts."*

Then, restate that valuable proposition. Show them reviews. Give them an offer, "Get $50 off your next DJ hire by referencing the coupon below. Call now."

When you set up your ad groups this way, you're going to have high relevancy, you're going to have a higher click-through ratio, and your conversions are going to be better because you're speaking directly to the consumers' needs. You are also giving them some type of call to action and maybe even a special incentive to choose you right at that moment.

If you are a wedding photographer and the person typed in any type of wedding photography term, then you need to have a specific AdGroup for that and a slightly different approach.

You can pick wedding photographer, best wedding photographer, party and event photographer etc.

> *"Capture the moments at your party, shower, corporate networking, or other celebrations. Visit my site to see my Portfolio, and call/email for a quote."*

Drive them to the wedding photography page on your website. On that page, speak to the specific search.

> *"We offer professional digital photography. From Weddings to Birthdays and just about any event, we will capture the sights of your special event in an unobtrusive Wedding Ceremony manner so that you can enjoy your event without having to deal with a pushy photographer."*

In some cases, you will be dealing with services that have a longer purchasing cycle.

Car repairing, for example. If it's general repairing or lift kit installation, if it's car hail damage repair, or even remodeling, customers are not necessarily going to pick up and call right at that moment.

They might just be in the researching process. For these types of services, offer them something for free, such as a free auto body inspection or 10 things to know before you remodel your existing car.

Have a lead capture form, where they can enter their name and their email address to download those guides.

This gives you the ability to catch people when they're in the first stages of their evaluation process. Send them a well-crafted guide that talks about what they should be thinking about and sets the buying criteria in your favor.

Educate and Engage

Why would they want to choose your company versus the competition? What things do they need to be made aware of? Do they need to make sure that they're dealing with somebody that is licensed and insured? Do they need to make sure that they're dealing with somebody who has vast experience in auto repairing, as opposed to just offering basic car repair service?

In that guide, you can really position yourself and educate them in a way that will make them want to utilize your services.

You can also use email marketing to send them messages over time. If they're at the beginning of a car repair or car remodeling project, you do your best to catch them early. Maybe it's going to be six months before they decide to make the final decision or to move forward with any type of project.

Because you got their email you could send them one email per week for the next six months. They're going to get something new from you once a week. Nothing annoying, but, "Here's an update, here's another thing, here's another interesting concept you can look at".

When they do get to the point that they are ready to move forward, they've seen you so many times and you've added so much value that they have no choice but to choose your company. You've made the decision easy for them.

This is a way to position yourself better for the longer purchase cycle projects, so you can capture more leads and convert them into customers.

These are many, many examples to illustrate the principals of how AdWords works. For each specialty in Law you would apply these same principals to your firm and how you approach your PPC campaigns.

AdWords Setup Best Practices

Here are some best practices when you get into Google AdWords (google.com/AdWords).

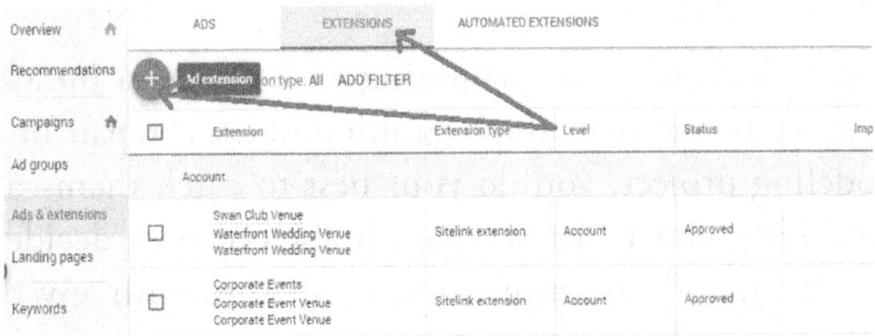

The first one is that you want to make sure you set up an extension with your address.

In chapter one, we showed you how to set up your Google Map listings and optimize it to rank in the Google Map. You want to use the same Gmail account that you claimed your map listing with on Google AdWords, so that you can come into extensions and add your address as an extension.

This gives you the ability to add your address and a direct link to your Google Places listing in your search.

In the screenshot above, you can see the ads for two Hail Damage Repair service providers. Not only do they have their text ad, but they've also got additional extensions on the ad with important site links, a link to their Google Places listing and their address. Can you see how that pops off these two ads a little more than the others? This is something you definitely want to consider. You want to take the time to set up your AdWords extensions.

These ad extensions allow your law firm to take up more 'real estate' on the web page and make your ad stand out even further!

Have Multiple Text Ads for Each AdGroup and Run Split Tests

The other best practice is to have multiple text ads for every one of your ad groups. This way, you can split test and see each of your ads and determine which one is converting better.

For example, split test by looking at these two ads for Steak Restaurant. They're just different variations of the same messaging. They're going to get equal share. If there are one thousand impressions, you could distribute 500 to one and 500 to the other.

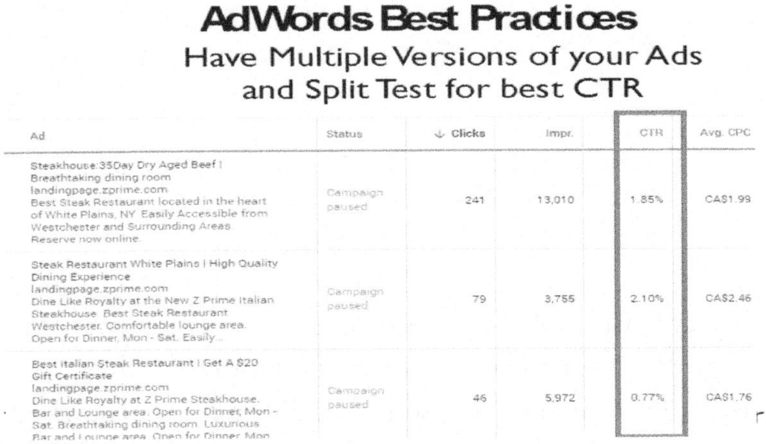

Ad	Status	Clicks	Impr.	CTR	Avg. CPC
Steakhouse:35Day Dry Aged Beef | Breathtaking dining room landingpage.zprime.com Best Steak Restaurant located in the heart of White Plains, NY. Easily Accessible from Westchester and Surrounding Areas. Reserve now online.	Campaign paused	241	13,010	1.85%	CA$1.99
Steak Restaurant White Plains | High Quality Dining Experience landingpage.zprime.com Dine Like Royalty at the New Z Prime Italian Steakhouse. Best Steak Restaurant Westchester. Comfortable lounge area. Open for Dinner, Mon - Sat. Easily...	Campaign paused	79	3,755	2.10%	CA$2.46
Best Italian Steak Restaurant | Get A $20 Gift Certificate landingpage.zprime.com Dine Like Royalty at Z Prime Steakhouse. Bar and Lounge area. Open for Dinner, Mon - Sat. Breathtaking dining room. Luxurious Bar and Lounge area. Open for Dinner Mon	Campaign paused	46	5,972	0.77%	CA$1.76

By split testing, you will be able to determine which one had a higher click-through rate. With that information, you can drop out the lower performing ad and create a new one.

Then at the end of the month, you can compare those two ads and see which one performed better. You keep doing that so you can continually improve your click-through ratios.

Remember, having better click-through rates is going to get you more traffic, but it's also going to give you a better-quality score. This will eventually make your cost-per-click lower, making it more profitable for you in the long-term.

Pay Attention to Average Position

Ad group	Status	Default Max. CPC	Clicks	Impr.	CTR	Avg. CPC	Cost	Avg. Pos.
Total - all ad groups			1,933	171,421	1.13%	$3.43		1.1
Social Media Marketing		$18.25	860	13,558	6.34%	$1.09		1.1
Email Marketing		$19.75	200	20,338	0.98%	$3.14		1.1
Website Design		$19.75	140	20,062	0.70%	$9.98		1.2
Web Design		$19.75	117	27,831	0.42%	$6.32		1.1
Social Media Advertising		$14.75	92	15,821	0.58%	$2.06		1.0

The other thing you want to do is to pay attention to your average position in your Google AdWords campaign. These settings are available making it very easy to analyze the data.

In the report above, you can see what position is being maintained. The average position is highlighted. This is based on the quality score and the average cost per click. You want to maintain a top four position on the major search engines in your pay per click marketing campaign.

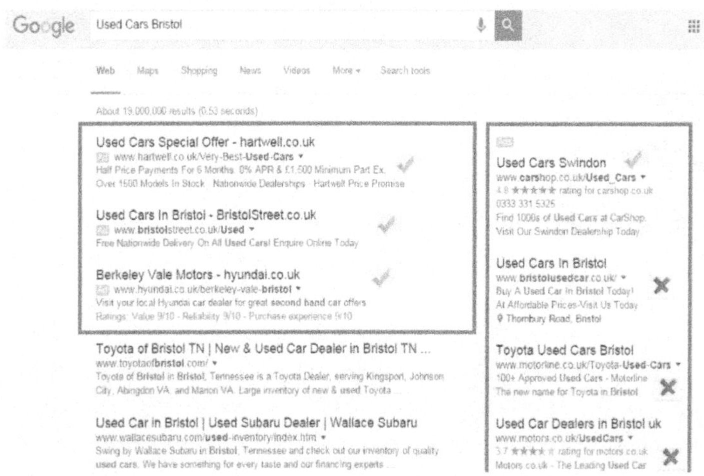

We have found that the further down the list they go, the higher the probability that you will be attracting a price shopper that's literally clicking every single company along the way.

You don't necessarily need to be the top listing, because that could just be a result of some random person that didn't think through what they're doing.

However, you want to maintain a top four position. That's going to give you the best overall visibility, and ultimately, the best return on your investment.

Pay attention to your average cost-per-click and manage your bids so you maintain a top four position.

Exact Match versus Broad Match

AdWords Best Practices

The other thing you want to pay attention to is exact match versus broad match.

You have a setting inside your AdWords campaign where you specify whether you want exact match or broad match.

Always Elect to do Exact Match

The reason is because if you choose broad match, you could very easily find yourself accidentally showing up on the search engines for a lot of keywords that have nothing to do with your specific business.

The other thing you want to do is pay attention to negative keywords – keywords that you don't want to show up for in the search engine.

A great example of this is jobs, employment, marketing, etc.

If someone types in "your city attorney," that's great. If they type in "your city attorney jobs," that's somebody looking for employment in the legal industry. Unless you are trying to fill a position or if you actually want to use your pay per click budget to get applicants, it's probably not the kind of the person you want to attract.

Setting up negative keywords means, for example, if someone types in "jobs," "employment," or "marketing services" anywhere in their search, it pulls you out of that search.

It pulls you out of that specific bidding process, so you won't be paying for clicks from somebody that's not relevant to you.

Mobile PPC

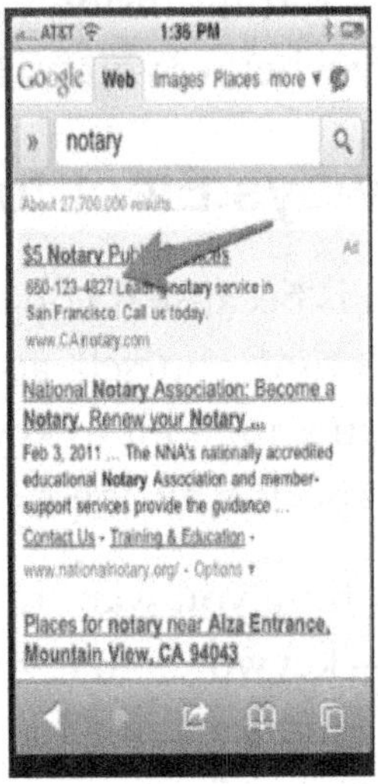

I talked a little bit about making sure that you've set up mobile pay per click campaigns. I've mentioned the major transition of people searching on their mobile device versus people searching on their computer.

Phone Searches versus Computer Searches

More and more people are accessing the Internet via smart devices: their iPhone, Android, and tablets. The searcher is typically in a different mind-frame when they are searching from a phone rather than from the computer.

When you're searching from a phone, you often just want to get the information right away, and/or want your problem solved as soon as possible. You can set up a campaign to have click-to-call built into your mobile campaign.

In the image above there was a search conducted from a mobile device, "Notary". Do you see the "Call" button towards the bottom? That's what we call a mobile PPC campaign with the click-to-call function turned on.

If somebody hits that "Call" button, they're connected immediately to that business. This is a quick alternative to having to search for the website and the phone number on your own. Plus, you can see on a mobile phone there is not a lot of screen space.

Those pay per click listings become really prominent and they dominate the search results page on mobile. A lot of times, you're going to get the majority of the clicks if you're in those top two positions. It's all about convenience, and the click-to-call function allows that.

It's extremely powerful to connect with these people that are searching from mobile devices. Set up a mobile-specific campaign and choose "Mobile Devices Only." Then you can pick your geolocation. That would be your 30-mile range or 20-mile radius. You then click a button to turn on the click-to-call function.

That's how you wind up with a pay per click campaign that has you in the top positions if you bid correctly, with the options for them to do a click-to-call.

Just to recap, you want to:

- Set up your ad groups correctly.
- Make sure that you pick keywords that group them together
- You write text ads that speak directly to that group of keywords, and
- Ensure your landing page (where you are sending those specific searches) speaks to the text ads and the group of keywords.
- You also want to be sure that you have some type of strong call-to-action that prompts your consumer into calling you as opposed to pressing the "Back" button and looking at four or five other competitors.

As the relevancy of your ad groups campaign and your keywords improve, your cost-per-click will decline and your conversion will improve.

You can spend less and still get better positioning and more traffic to your website. This is how you maximize the profitability of your pay-per-click marketing campaigns and succeed in PPC where others fail.

FOURTEEN

Paid Online Directories -

What Paid Online Directories Should You Consider Advertising in (Yelp, Foursquare, Yellow Pages, Better Business Bureau, Merchant Circle, Etc.)

In this chapter, we're going to be covering paid online directory listings.

We talked about the overall Internet market strategy, beginning with the foundation of having a properly optimized website. We have also discussed making sure that you've got yourself set up with all the right pages on your website, the conversion elements, doing the off page optimization for building inbound links, building authority for your domain, having the review acquisition strategy, and making sure that you're ranking in the organic, non-pay-per-click listings for your most important keywords.

We then talked about looking at social media and email marketing as a way to connect with your customer on a deeper level and get more repeated referral business. As you get those non-paid elements of your Internet marketing strategy squared away, you can start looking at paid online marketing programs.

We talked about pay-per-click marketing, and the way you could set up an effective pay-per-click marketing campaign on AdWords or Microsoft Bing search in order to show up in the paid listings.

In this chapter, I want to talk about other paid marketing components, such as online directory listings that you can pay for to get premium listings.

There are literally hundreds of online directories, from Yelp.com and Foursquare, to City Search, and BBB, as well as an array of smaller secondary directories. I'm going to talk about the ones that are the biggest; the ones that will help you gain exposure where your customers are looking most.

Paid Online Directory Listings and Online Sites You Should Consider

As mentioned, there are literally hundreds of online directory listings. The ones that we have found to be the most prominent and visited are:

- Yelp
- Yellow Pages
- Foursquare
- BBB
- Merchant Circl

If you have an unlimited budget and you are already doing well with your organics, and you wanted to pay for some additional premium placement in online directories, these are the ones I would suggest that you take into consideration.

Yelp.com

Yelp is one of the best review sites for local businesses. If you want customers to find your business online, you need to be on Yelp. Yelp allows you to send public or private messages (including deals) to customers and review business trends using the Yelp reporting tool. If, for some reason, you have a lot of reviews on Yelp.com, it might not be a bad idea to pay for a premium ad on their directory for your services.

YP.com

The online yellow pages vary area by area. In some markets, it is YP.com and in others, it is DexKnows.com, Version Yellow Pages, YellowBook.com or some similar version.

With YP.com and other online yellow pages, you need to be very careful when you get started. You don't want to be roped into their print Yellow Page ad. The cost goes from a couple hundred bucks a month to potentially a couple thousand dollars per month when you start to get into their Yellow Pages book and their pay-per-click advertising.

Do not let Yellow Pages manage your pay-per-click advertising under any circumstances. There is a whole chapter on Pay-Per-Click Marketing in this book. I go into great depth about how to set up an effective pay-per-click campaign. You don't want to let any of these companies try and touch your pay-per-click advertising on Google, Yahoo or Bing. Do it the right way. Set up the ad groups on your own.

City Search

City Search specializes in listings for restaurants, bars, spas, hotels, restaurants, and other businesses across the U.S., optimizing them via a partner network that includes Expedia and MerchantCircle.

Better Business Bureau

BBB, the Better Business Bureau is not just an online directory, it is a major sign of credibility. It's not as popular as it once was but posting the BBB logo and being able to say that you're A+ credited is worth the investment. I haven't found that it allows for a ton of leads, but it's a great credibility symbol and a good thing to be able to reference.

FOURSQUARE for Business

Foursquare

Foursquare is nowhere near as popular as Yelp, but it does provide listings for all kinds of local businesses. 93 percent of local storefronts represent 2 million of the businesses who are already listed on Foursquare, and the site is visited by more than 50 million people.

mc merchantcircle.com

MerchantCircle

MerchantCircle allows people to find the best local merchants. The site includes listings for all kinds of merchants and business owners, ranging from attorneys and notaries to realtors and agencies. Over 100 million consumers visited the site last year to search its listings of 2 million businesses. The site gets around 340,000 monthly visits.

Pay-Per-Lead and Lead Services - How to Properly Manage Pay-Per-Lead Services for Maximum Return and Long-term Gains

Now let's talk about pay-per-lead services.

With these services, you can pay per lead or you can pay on a per-monthly basis to gain access to all the leads that come into your market. I am not saying you should 100% do this. I am simply suggesting that if you need some additional leads or you've got an inside sales team that can follow up with these proactively, these are some good options.

- Emfluence.com
- Fuellead.com
- Intellibright.com

There is an abundance of these types of services. The best way to find additional lead services specific to your business would be to run a Google Search for "Your Service Lead Service", "Your Service Pay-Per-Lead", etc.

How Do Pay-Per-Lead Services Work?

The nice thing about this type of service is that you only pay when you get a qualified lead. With others, you just have a budget. For example, you set $500.00/month to get all of the leads that come in from that area. Most of these pay-per-lead service providers have a combined experience in working with a wide range of industries including automotive, law, healthcare and other industries.

If you have followed the plan outlined in this book, you should have your organic keywords ranking well in the search engines and map listings, proactive social media and email marketing as well as a well-structured pay-per-click marketing campaign. If you want to bump the lead flow, these services can help to start channeling new people that are in the market for your services.

However, you must be diligent and quick with your follow up.

You will hear a lot of horror stories about how badly these lead services work and how you can throw so much money away. I will be the first to say that I don't think it's the place to start. If you have built your Internet marketing strategy on pay-per-lead services, you're destined to fail. You can't build a sustainable business around just this one strategy.

But, if it's an addon to a strong Internet marketing program, then it can be relatively effective. The key is to remember that these requests for leads aren't coming to you directly. They're on Emfluence. They're on Fuellead.

They are sending in an anonymous request for a quote,

providing their name and email address knowing that they are going to get phone calls. However, they are probably going to be price conscious shoppers. They are using these services because they want to get the lowest price possible. Keep that in mind.

Fast Follow Up Is Critical

If you don't have the time and energy to chase leads, then I would say to pass on pay-per-lead services altogether. These leads also go out to you and a number of other law firms in your area, so you have to be aggressive. You must be the first person to get customers on the phone and you have to be professional with a compelling offer that makes them want to choose you as opposed to the competition.

You also need to create a followup system to make sure that you have a fallback plan in place for leads that you can't reach right away. You can get these leads in a variety of formats. They'll send you an email, you can log in and download an Excel list, or you can receive a text message that alerts you as soon as the email comes through.

If you have a marketing manager on your team, be sure to assign somebody specifically to follow up on leads. Know who is accountable for these leads when they come in.

If it's going to you, to your paralegal, or even one of your attorneys, you don't want there to be any confusion about who is responsible for following up because then the lead falls through the cracks.

Specifically assign someone the responsibility of reaching out to these people. Have a predefined script on how the call should be handled. Be professional. Be courteous. Be quick.

A lot of these are going to go to the first person that gets them on the phone, so it is important to be aggressive. Don't just call once. Have a process in place where you reach out to these people 3 to 5 times over the course of the next 24 hours because they're in the window to buy.

Then, have a fallback strategy, in the event that you don't get them on the phone. If you don't get them on the line, make sure that you're taking note of their name and their email address so that you can remain top-of-mind with them.

The reality is this is somebody in your service area that is in need of your specialty.

If you're not sending an email follow-up, and if you're not adding them to your email-marketing database, then you're wasting marketing dollars. If you've just spent $5, $10, $25 for that lead and you're not proactively and diligently following up with them via email, you might as well not even pay for this service.

Below is a script of a solid fallback strategy.

Set up an email auto-responder on a program such as Aweber or Mailchimp, where your marketing manager can enter the customer's name and email address and have a series of emails that go out to the customer over the next several days.

Remember not to let this be your crutch. Don't think that these emails are going to do the trick.

Email 1 – Subject – Your Recent XYZ Service Inquiry

Customer Name,

You recently submitted a request on [LEAD Site] for help with XYZ Services. I called and left a message for you on the number that was listed and look forward to talking with you soon. You can reach me directly at xxx-xxx-xxxx. With so many XYZ companies to choose from in [YOUR CITY], I know it can be hard to know who you can trust.

At XYZ Company we have been serving the [your city] area since 1982 and are dedicated to resolving your XYZ issue quickly and cost effectively. Give me a call at xxx-xxx-xxxx to schedule your service.

Email 2 – Special Offer for XYZ Services

Customer Name,

You indicated that you were in some need of some XYZ services a few days ago. I'm sure you have received a number of calls from XYZ Company, who are eager to earn your business.

WELL – as our outside-of-the-box approach to getting your attention, we want to offer you a special offer. If you call us today and reference this coupon, we will do a free consultation.

<ATTACH COUPON IMAGE>

Call now and get free consultation with XYZ Company.

Email 3 – Subject – RE: Your Recent Business Inquiry

Customer Name,

You reached out to us earlier this week via [lead site] looking for some help with your XYZ service. We would love to be of service to you. I have tried you a few times on the phone number you listed with no success and don't know if you are just busy or if you already hired another law firm. Please shoot me a quick reply to let me know if we can be of assistance or give me a call at xxx-xxx-xxxx.

The aggressive follow up work on the phone is what's going to get you the business. So just have this as a fallback strategy.

Stay in Touch

Again, don't stop there. You've got their name and email address. You should be marketing to these people via email on at least a monthly basis. You should have an email database of customers and prospects that you should be sending out emails to once a month with some type of update.

"Here's what's going on with our law firm. Here's why you should consider our custom services". Include some special offer incentive." This is to remain top-of-mind so that you can build your customer-base both in email and social media.

As you look at paid online advertising and paid-per-lead services, be cautious. Don't overspend. Put the tracking in place to make sure you've got a strong return on investment. If you are going to play the pay-per-lead service game, make sure that you have a proactive, diligent process that touches these people multiple times, via phone and email.

FIFTEEN

Track, Measure AND Quantify -

How to Track Your Online Marketing Plan to Ensure That Your Investment is Generating a Strong Return on Investment

Congratulations! Now That You Have…

- Built and optimized your website
- An ongoing link building strategy in place where you're creating inbound links and moving up in the search engines
- Implemented email marketing and social media marketing initiatives, and
- Possibly implemented a paid online marketing campaign including Pay-Per-Click and Pay-Per-Lead services...

…You need to put some tools in place so that you can track, measure and quantify your data to ensure that you're moving in a positive direction.

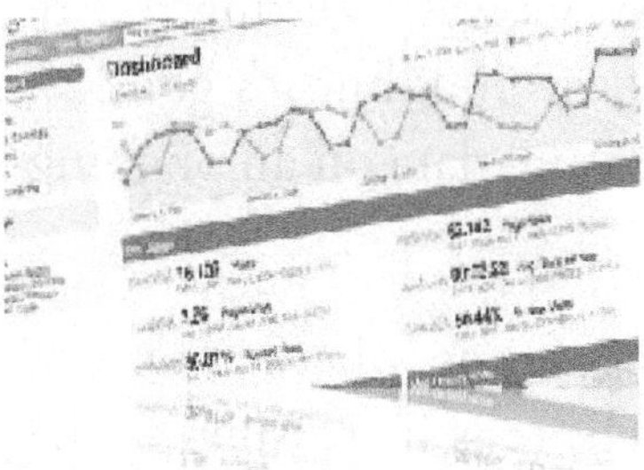

- Google Analytics www.google.com/analytics
- Keyword Tracking Report www.gshiftlabs.com
- Call Tracking Report - www.callfire.com

Analytics Tracking

There are a lot of different tracking mechanisms that you can put in place. I'm going to recommend three core tracking mechanisms:

- Google Analytics
- Keyword Tracking

- Call Tracking

The first is Google Analytics. Google Analytics is a great website data analysis tool and it's completely free. Google Analytics will show you specifically:

- How many visitors got to your website on a daily, weekly, monthly, and annual basis
- What keywords they typed in to get there
- What pages on your website they visited
- How long they stayed

The main thing you want to see from Google Analytics is where you started and where you are now.

You want to ask yourself: When I started this whole Internet marketing process, how many visitors was I getting to my website? Maybe it was 5, 20, 100, or 500, but it's good to know. Then you can compare to future data on an ongoing basis.

Ultimately, what you are looking for is whether or not the number of visitors to your website is increasing. Is the variety of keywords that they're finding you with increasing? Are you moving in a positive direction?

You can also set up reports within Google Analytics. To get set up on Google Analytics, you just go to Google.com/analytics. It's a simple process. You verify that you own the website through a variety of different methods, and then install a small piece of code into your website's HTML. After you have done that, you've got the tracking in place and are ready to go.

Sample Google Analytics Traffic Report

Traffic Overview:

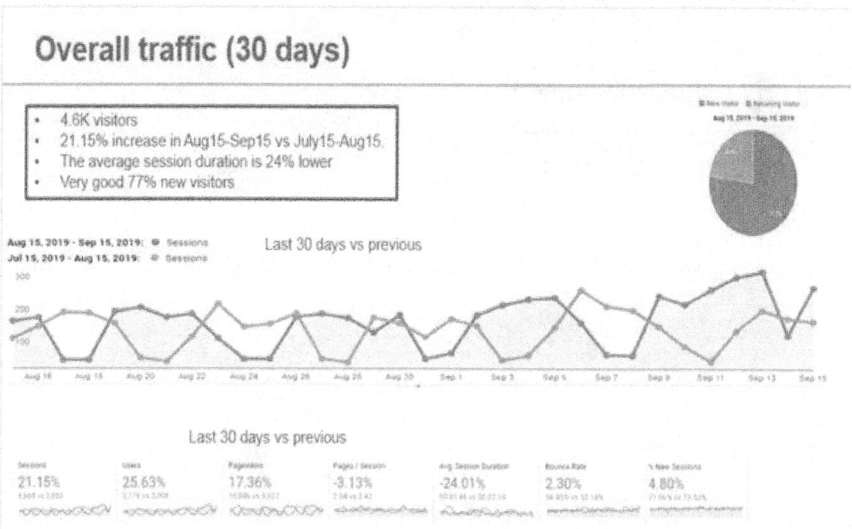

Traffic by Page:

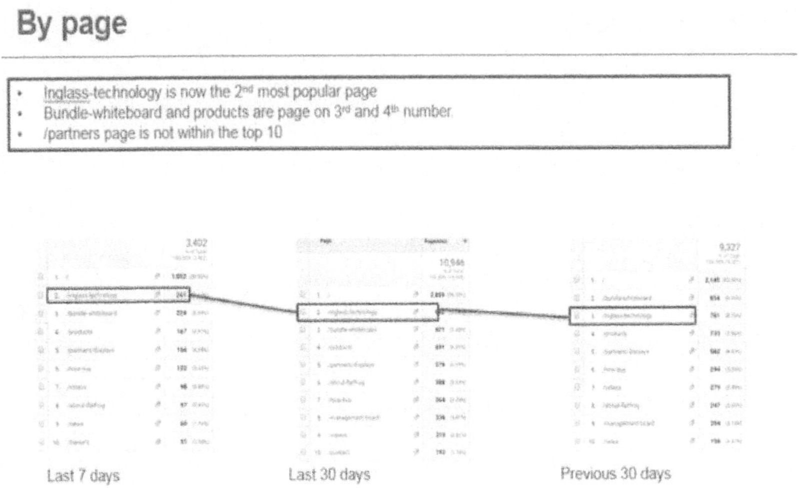

Traffic by Location:

Traffic Flow:

Web flow (30 days)

4 path types
1. Software
2. Hardwar
3. Bundle
4. Random

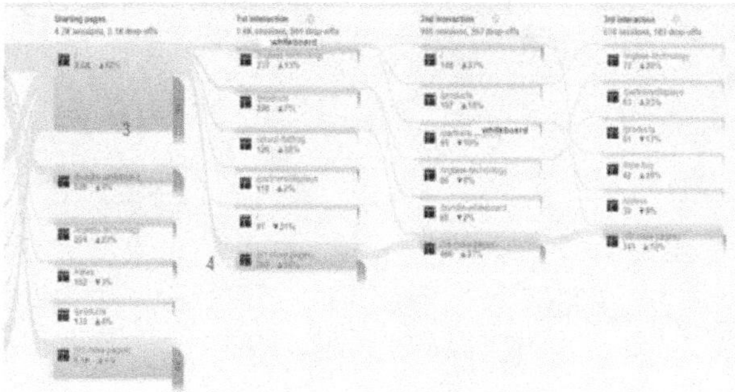

Traffic by Device:

By device

- 81% of the media is desktop
- 59% of desktop is Chrome
- 59% of mobile is iOS

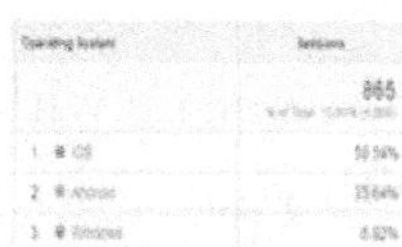

Demographics:

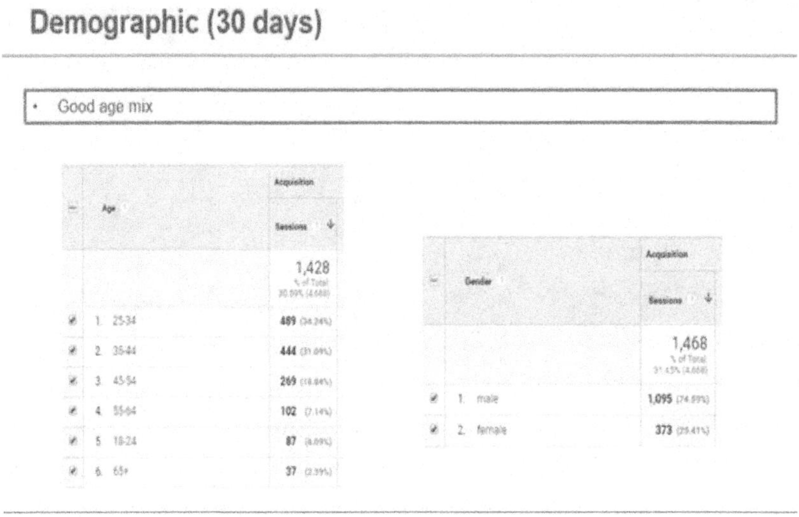

Keyword Tracking

The other tracking mechanism that I recommend is keyword tracking.

At the beginning of this process, we talked about keyword research to determine what keywords people are typing in when they need your services.

We came up with a list and all those keywords were combined with your cities and sub-cities.

There are tools that will tell you how you're ranking on Google, Yahoo, and Bing for those various keywords. A few options include:

- Bright Local
- White Spark
- Raven Tools
- WebCEO

The keyword tracking tool I recommend is called BrightLocal.

You can learn more about it at www.brightlocal.com. There is a cost associated with this service, but it is great resource for tracking your search engine optimization progress.

You take your keywords, put them into the BrightLocal Keyword Tracker and then set up a weekly and monthly report that shows where you rank on Google, Yahoo and Bing for your most important keywords.

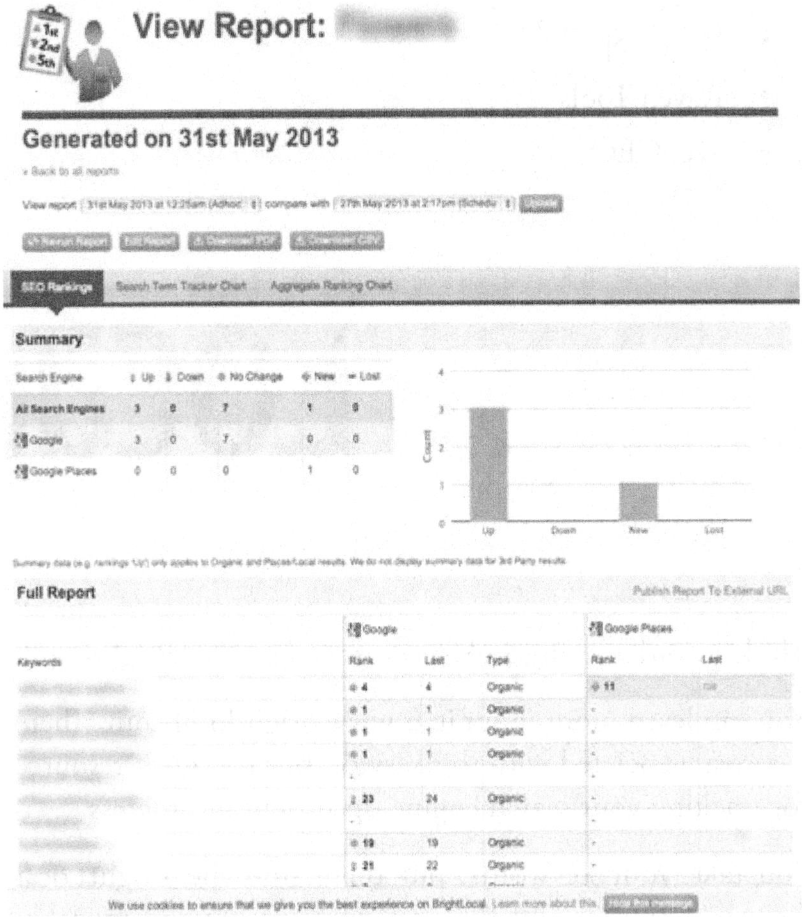

With a report like this, you can easily see how your website is trending in the search engines.

You'll see yourself move up in the results if you've built out the website correctly with the right on-page factors (title tags, H1 tags, meta descriptions, etc.), if you're building links, developing citations and have a proactive review acquisition system in place.

If you see yourself stagnating, you can go back to that keyword, figure out which page is optimized for it, look at your links and link profile, and whatever is necessary to push that keyword to the next level.

Call Tracking

The third really important tracking mechanism that I recommend is call tracking. Having better rankings and more visits to your website is all fine and dandy, but in most businesses, nothing happens until a call is made.

Calls are crucial to your business. You want to have some type of tracking mechanism in place to know how many calls are coming in on a monthly basis and what's happening within those conversations.

Are calls turning into sales? That's where the rubber meets the road. That's why we're doing all of this. Who cares if you're in the number one position if it doesn't result in dollars to the business?

There are several call tracking tools that you can use. Here are a few:

- CallRail
- DialogTech
- CallSource
- Century Interactive
- CallFire

One of the tools I've seen used prevalently is called CallFire. You can learn more about it at www.CallRail.com.

Most of these call-tracking services will let you choose a phone number based on your area code. So, you type in the number you want to get. It's a nominal fee on a monthly basis ($2 - $5 per month), and you get a tracking number. CallRail currently starts at $45/month but that is for 10 lines and they have very good programs, tracking, reports and call recording.

Then, you can take that tracking phone number and you can put it on the graphics on your website so that you can track the number of calls and even listen to recordings of the conversation.

That number will be set to ring in your office. It's just a forwarding number. If somebody dials it, it still rings to your office like always, but it is a tracking number.

You can report on the number of calls using the Internet and play back recordings of those conversations. It's extremely powerful to know the number of calls you were getting when you started versus the number after you incorporated your new marketing strategy.

You can go in and listen to those conversations and ascertain how many of those calls turned into booked service while knowing what the revenue associated with that service is. That is how you get a true gauge on the return on investment associated with your online marketing strategy.

These are the types of tracking mechanisms I recommend. There are a lot of different things you can do, but having analytics, keyword tracking, and call tracking really gives you the most important key performance indicators to gauge your progress.

SIXTEEN

Next Steps

Throughout the course of this book, we have covered an abundance of information.

We've mapped out your Internet marketing plan and taken you step-by-step through how to claim and optimize your Google map listing, how to optimize your website for the most commonly searched keywords in your area and how to leverage social media to get more repeat and referral business.

We then covered paid online marketing strategies like pay-per-click and pay-per-lead services. If you have taken action and followed our instructions, you should be well on your way to dominating the search engines for the keywords in your area.

Need More Help?

If you've gotten to this point and feel like you need some extra help to implement these ideas, we are here to support you. As experts in helping online businesses across the nation, we have had tremendous success implementing these strategies.

You can call us directly at 404-777-4041 with any questions that you might have. Our team will review your entire online marketing effort (Website, Competition, Search Engine Placement, Social Media, etc.) and come back to you with a complete assessment of how you can improve and what you can do to take your online marketing efforts to the next level.

Request A Free Custom Online Marketing Evaluation Now

Your Custom-Tailored Optimization Audit will:

- **Identify** key issues that could be harming your website without you even knowing it.

- **Look** at where your website stands compared to your competitors.
- **Determine** whether SEO is the appropriate route for you to take.
- **Uncover** hidden revenue that you're leaving on the table.
- **Offer** recommendations that you can put to use immediately.

Schedule your custom audit at

www.LawFirmMarketingServices.com

ABOUT THE AUTHOR

Kenneth Sullender began in the seo industry in 2000 and over the years has built dozens of digital assets and apps. many of these have been bought and sold to private and public organizations with valuations over 7 figures.

His desire to understand seo, social media, ppc, algorithms, and online digital marketing was conceived out of the necessity to grow and manage these assets.

Through decades of working and becoming an expert in seo and online digital assets, other business owners would reach out for consulting services. After witnessing so many others being over charged and underserved by 'seo experts', he launched internet marketing pros and subsequently law firm marketing services specializing exclusively in law firm seo and digital marketing.

Over the years, the team has provided digital online services with integrity that are transparent and results driven to many law firms.

Made in the USA
Monee, IL
13 September 2026

62976622R00148